应用型高等院校校企合作创新示范教材

Java Web 开发技术与项目实战

主　编　罗如为

副主编　陈镇铖　武佩文　张志昊

中国水利水电出版社
www.waterpub.com.cn
·北京·

内 容 提 要

目前，软件企业开发基于企业级 JavaEE 的软件项目，都会选择一种框架技术或几种框架技术的整合，如 Nutz、SSH、SSM、JFinal，选择框架技术最主要的目的是提高软件项目开发效率，所以掌握一种主流框架技术是很多企业对开发人员的基本要求。本书针对 JavaEE 软件工程师岗位的特点，全面创新本书的结构体系，努力体现"项目任务驱动"教学和"案例"教学相融合的课程特色。全书共 7 章，第 1 章和第 2 章作为项目实战前的准备篇，介绍了开发资料的下载、开发环境的搭建、系统的需求分析和设计、数据库的设计；第 3 章至第 6 章作为框架技术项目实战篇，每一章介绍了一种框架技术或框架技术组合的开发应用；第 7 章介绍将项目脱离开发环境，部署到服务器运行。

本书主要通过项目实战，用框架技术实现"新闻发布系统"。项目案例中包括了 Ajax、Beetl 模板引擎、EasyUI 前端框架、在线 HTML 编辑器 UEditor 等前端技术的应用。

本书图文并茂、深入浅出、语言流畅，强调工程思想。书中包含大量精心设计并调试通过的编程实例，方便初学者学习。例如，书中讲解基于 Nutz 框架技术，采用了软件企业常用的快速迭代方法实现系统开发，让学习者提前感知软件开发人员的工作。

本书特别适合作为高等院校的学生作为学习 JavaEE 框架技术开发 Web 项目相关课程的教材和参考书，也可供从事 JavaEE 应用系统开发的人员参考。

图书在版编目（CIP）数据

Java Web开发技术与项目实战 / 罗如为主编. -- 北京 : 中国水利水电出版社, 2019.2
应用型高等院校校企合作创新示范教材
ISBN 978-7-5170-7446-5

Ⅰ. ①J… Ⅱ. ①罗… Ⅲ. ①JAVA语言－程序设计－高等学校－教材 Ⅳ. ①TP312.8

中国版本图书馆CIP数据核字(2019)第031185号

策划编辑：周益丹　　责任编辑：周益丹　　封面设计：梁　燕

书　　名	应用型高等院校校企合作创新示范教材 Java Web 开发技术与项目实战 Java Web KAIFA JISHU YU XIANGMU SHIZHAN
作　　者	主　编　罗如为 副主编　陈镇铖　武佩文　张志昊
出版发行	中国水利水电出版社 （北京市海淀区玉渊潭南路 1 号 D 座　100038） 网址：www.waterpub.com.cn E-mail：mchannel@263.net（万水） sales@waterpub.com.cn 电话：（010）68367658（营销中心）、82562819（万水）
经　　售	全国各地新华书店和相关出版物销售网点
排　　版	北京万水电子信息有限公司
印　　刷	三河市鑫金马印装有限公司
规　　格	184mm×260mm　16 开本　13 印张　314 千字
版　　次	2019 年 2 月第 1 版　2019 年 2 月第 1 次印刷
印　　数	0001—3000 册
定　　价	36.00 元

前　言

JavaEE是企业级Web应用开发中的一种软件开发技术，它与企事业单位的需求联系密切，且不断被改进，不断融入新的思想和新的解决方案。掌握JavaEE开发技术的人员，就业地域广、选择多、薪资高，因此这一技术被广大Java爱好者和软件公司所青睐，同时使得“JavaEE框架技术”课程成为高校计算机相关专业的一门主要专业课。

本书由高校专业教师与企业软件工程师合作编写，作者的软件开发经验均在9年以上，具有丰富的教学与实践经验。本书针对应用型本科院校学生培养定位，既强调基本知识的理解，更注重基本技能和工程应用能力的培养，使学生了解企业对软件开发人才的实际需求，拓宽学生的知识面，掌握开发Java Web项目流程，具备独立开发项目的实践能力，提高在软件开发过程中发现问题、分析原因、解决问题的能力，激发学生的学习兴趣，创新性地开发自己感兴趣的Web应用系统。

本书以新闻发布系统案例为实战项目。首先介绍了开发环境与工具；然后介绍新闻发布系统的需求分析，并根据需求来设计系统和数据库；接下来介绍了如何应用Nutz、SSH、SSM和JFinal四种不同框架技术开发新闻发布系统；最后将开发的系统部署到服务器。每一部分都设计了若干有针对性的考核任务，每个任务包括若干考核要点；考核任务按照知识点进行设计，循序渐进、逐步深入，将理论知识学习与实践能力训练融为一体，同步进行。如果每一个项目的阶段性任务都完成了，也就完成了一个项目的完整开发。

本书中所介绍的项目均在Windows 7、MyEclipse2015、JDK1.8、Tomcat8、MySQL5.7环境下进行开发，使用的后端框架为Nutz 1.r.65、MyEclipse2015集成的Struts 2.2.1+Spring 4.1+Hibenat 4.1、MyBatis 3.4.6，使用的前端框架主要是EasyUI，在线编辑器使用百度官方提供的UEditor 1.5.0。每个项目案例已开发实现，并调试运行，功能正常。同时也给出了完整的实现步骤，从Web项目系统的设计到系统的部署，读者按照书中所讲述的内容实施，可以顺利地完成开发任务。

学习软件开发技术，无论是后端还是前端，都不宜拘泥于某一种技术本身，最重要的是学会解决问题的方法，掌握了方法，无论用什么技术都可以较快地开发出软件系统。对于初学者，没有捷径可走，需要在不断地编码、运行、调试过程中总结经验。遇到问题时，首先要分析问题，确定问题的来源，然后充分利用搜索引擎、使用手册、说明文档和身边的老师与同学，找出解决问题的方法，不断地积累解决问题的经验，高质量地完成任务，最终掌握软件开发技术、学以致用。

本书由罗如为任主编，陈镇铖、武佩文、张志昊任副主编。书中第3章由陈镇铖、武佩文和张志昊编写，各项目的前端页面内容由张志昊编写，其他内容由罗如为编写，全书由罗如为负责审核和统稿。羊四清老师对本书的编写提出了很多宝贵意见，在此，向他表示感谢。

由于编者水平有限，书中难免有疏漏之处，敬请广大读者批评指正。

编　者

2018年11月

目　　录

第 1 章　开发环境与工具

1.1　下载资源

建议从官方网站下载本书使用的开发工具，如 JDK、Tomcat、MySQL、MyEclipse、HeidiSQL，下载链接地址参照附录，推荐优先下载免安装版本（Portable）。

1.2　安装 JTM

JTM 是指 JDK + Tomcat + MySQL 集成环境。安装和配置好 JTM，就能搭建出支持 JavaEE 项目和 MySQL 数据库的服务器运行环境，可用于 JavaEE 项目的测试和部署。

环境安装可选择 exe 或 msi 的运行安装版本，这样可以通过安装程序的向导方式，根据提示一步步完成安装，大部分选择默认安装项，但默认安装的路径可能存在空格或中文，在调试程序时可能遇到因路径问题引起的特殊异常。

为了避免遇到安装路径引起的异常，也为了便于管理开发部署的项目，可以选择解压缩版本，将 JTM 安装到同一目录，目录创建的注意事项如下所述：

（1）目录名尽量短小，但应见名知义。

（2）目录名中，尽量使用小写字母或字母与数字的组合，建议不要使用中文或其他符号，尤其不要使用空格。

（3）建议把目录创建在非系统盘中。

假定 JTM 目录为 D:\jtmlrw，则主要目录结构如图 1-1 所示，当然还有其他安装后的文件夹或文件，一般不要删除。

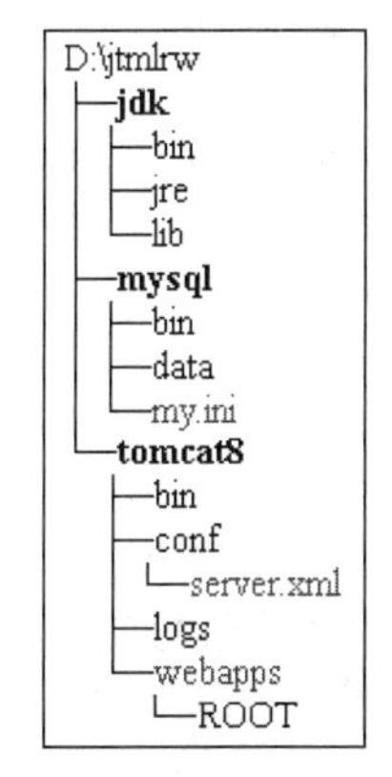

图 1-1　JTM 安装后的主要目录结构

1.2.1　安装 JDK

将下载的 jdk1.8 压缩包解压到 D:\jtmlrw 目录下，然后设置 3 个环境变量：

（1）新建系统变量 JAVA_HOME，设置值为 D:\jtmlrw\jdk。

（2）新建系统变量 CLASSPATH，设置值为.;D:\jtmlrw\jdk\lib;D:\jtmlrw\jdk\lib\tools.jar。

（3）在系统环境变量 path 的开头，添加 D:\jtmlrw\jdk\bin;D:\jtmlrw\jdk\jre\bin;。

一定要注意是添加，不要覆盖原来的 path 值。

1.2.2　安装 Tomcat

将下载的 Tomcat8.0 压缩包解压到 D:\jtmlrw\目录下，设置 URI 默认字符集为 UTF-8，可以根据需要修改默认的 8080 端口。

打开配置文件 tomcat8\conf\server.xml，在 8080 端口所属的 Connector 节点，添加 URIEncoding="UTF-8"，如图 1-2 所示，可以解决大部分 GET 请求时中文乱码的问题。

★非常重要★

```
<Connector connectionTimeout="20000" port="8080" protocol=
"HTTP/1.1" redirectPort="8443" URIEncoding="UTF-8"/>
```

图 1-2 Tomcat 设置 URI 默认字符集为 UTF-8

1.2.3 安装 MySQL

（1）将下载的 MySQL5.7 压缩包解压到 D:\jtmlrw\目录下，设置环境变量。

在系统环境变量 path 的开头，添加 D:\jtmlrw\mysql\bin;

（2）由于 5.7.20 以上的 MySQL 解压缩版文件夹已没有默认的 data 文件夹和配置文件 my.ini，所以需要先在 D:\jtmlrw\mysql 目录下创建 my.ini，添加必要的配置项，如下所示。如果有更多的配置需求，可以查看官网提供的文档。

```
[client]
port=3066
[mysql]
default-character-set=utf8
[mysqld]
port=3066
datadir=D:/jtmlrw/mysql/data
character-set-server=utf8
default-storage-engine=INNODB
sql-mode="STRICT_TRANS_TABLES,NO_AUTO_CREATE_USER,NO_ENGINE_SUBSTITUTION"
server-id=1
max_connections=151
table_open_cache=2000
tmp_table_size=16M
thread_cache_size=10
key_buffer_size=8M
read_buffer_size=0
read_rnd_buffer_size=0
max_allowed_packet=40M
sort_buffer_size=256K
table_definition_cache=1400
```

生成 data 文件夹及默认的 MySQL 数据库，在命令行窗口中运行“mysqld --initialize-insecure --user=mysql;”命令。

（3）将 MySQL 设定为系统服务，随开机启动。启动管理员模式下的 CMD，运行“mysqld –install MySQL;”命令；启动 MySQL 服务，运行“net start MySQL”命令。

1.2.4 安装问题

不同的计算机，系统环境有差异，安装过程中可能会遇到一些问题，例如：

（1）安装目录中有空格，或者使用了中文。

（2）端口号冲突。

解决方法（以 MySQL 为例）：

- 可用命令“netstat –ano”找到占用端口的进程 pid，如图 1-3 所示，找到占用 3066 端口的进程 pid 为 2032。不同时间、不同机器所显示的 pid 不一定相同。

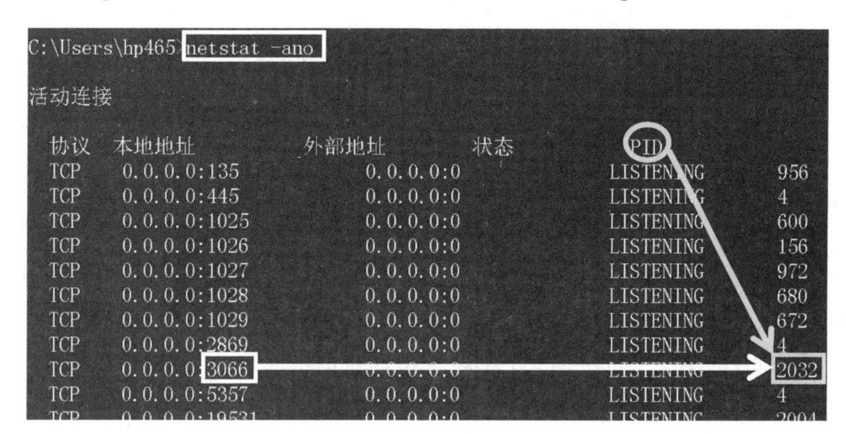

图 1-3　用命令查看进程占用的端口列表

- 在任务管理器中，根据上一步找到的 pid 确定占用端口的进程，如图 1-4 所示，可以看到 pid 为 2032 的进程是 mysqld.exe。

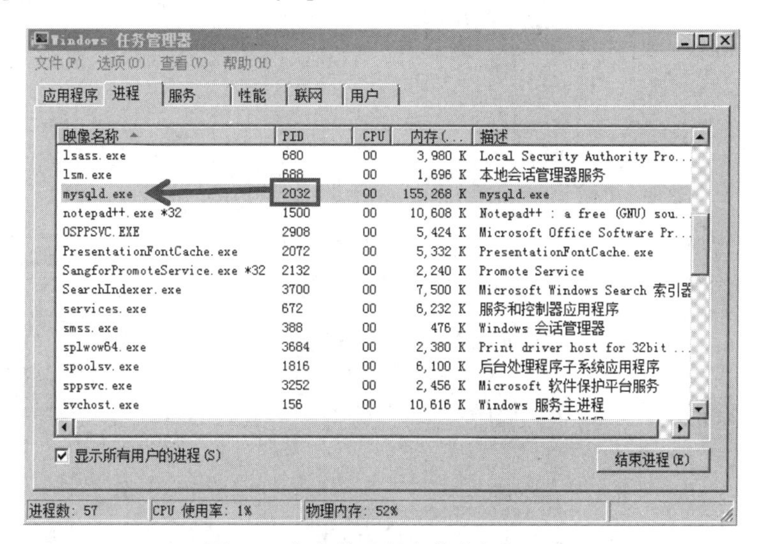

图 1-4　任务管理器中的进程与 pid

- 根据实际情况卸载占用端口的进程所对应的软件，或者更改 JTM 的端口号。

（3）安装 mysql 5.7.xx 报错，提示信息为：“This application requires Visual Studio 2013 Redistributable. Please install the Redistributable then run this installer again”。

问题原因是 MySQL 自动安装时所需的 Visual C++ Redistributable 路径不正确，或者 x64 的 MySQL 识别的也是 x86 的安装路径，所以解决方案是从微软公司的官方网站上手动下载 Visual C++ Redistributable（https://www.microsoft.com/zh-CN/download/details.aspx?id=40784）

安装，将 vcredist_x64 和 vcredist_86 都下载下来，先安装 vcredist_x64 再尝试重新安装 MySQL 5.7.xx。如果依然报错，则安装 vcredist_x86 后再次安装 MySQL 来解决问题。

1.3 安装 MyEclipse

1.3.1 安装

从官网上下载 myeclipse-2015-stable-3.0 运行安装，安装路径（如 D:\MyEclipse2015）和工作空间 workspace（如 F:\myjavaee）尽量选择（或创建）一级目录，目录名尽量短小，使用英文或数字。

1.3.2 常规设置

使用 MyEclipse 开发调试项目前，需要进行参数设置。运行 MyEclipse，单击 Windows 菜单中的菜单项 Preferences（首选项），打开参数设置对话框，以下参数的设置均在 Preferences 对话框中进行。

（1）单击 General 中的 Workspace，如图 1-5 所示，设置 MyEclipse 工作空间中项目文件（*.java、*.js、*.css、*.json 等）的统一编码方式（Encoding）为 UTF-8。**★非常重要★**

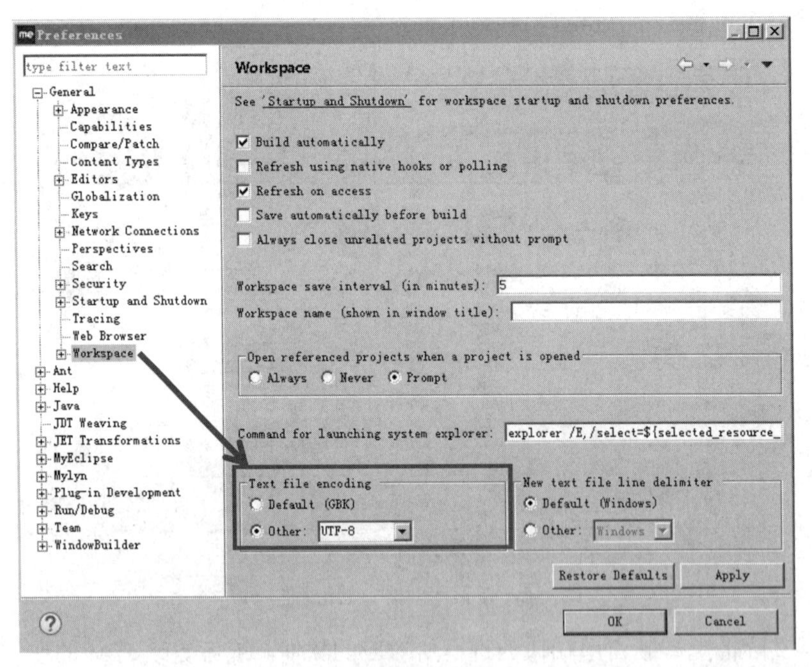

图 1-5 设置项目文件编码方式

（2）单击“+”号，展开 General 和 Appearance，选择 Colors and Fonts，字号设置稍大一点，如“14”。

（3）选择 General 中的 Web Browser，选用系统默认的外部浏览器，通常选择便于调试的极速模式浏览器，例如 360 极速浏览器（http://chrome.360.cn/），如图 1-6 所示。

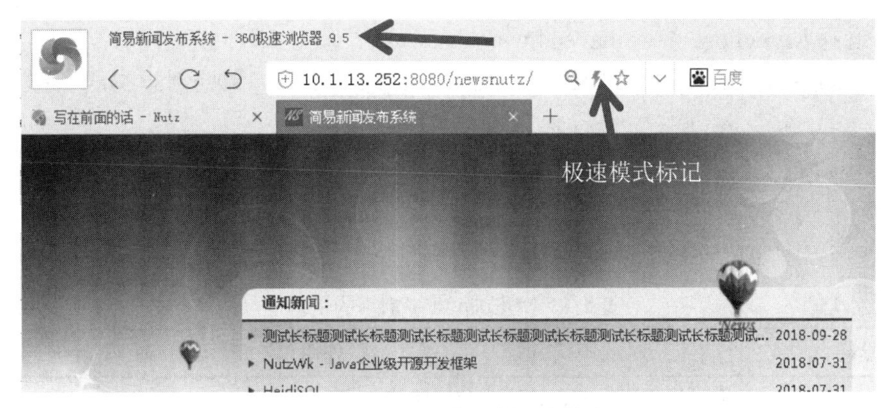

图 1-6　设置浏览器为极速模式

（4）选择 Java 中的 Installed JREs，添加已安装的 jdk，如 jdk1.8。

（5）展开 MyEclipse 中的 Files and Editors，设置各类文件的 Encoding 为 UTF-8

（6）也可以搜索需要设置的项，如搜索“server”，选择 Runtime Environments，设置调试服务器，如图 1-7 所示，设置 runtime 为已安装的 Tomcat8，JRE 为已安装的 jdk。

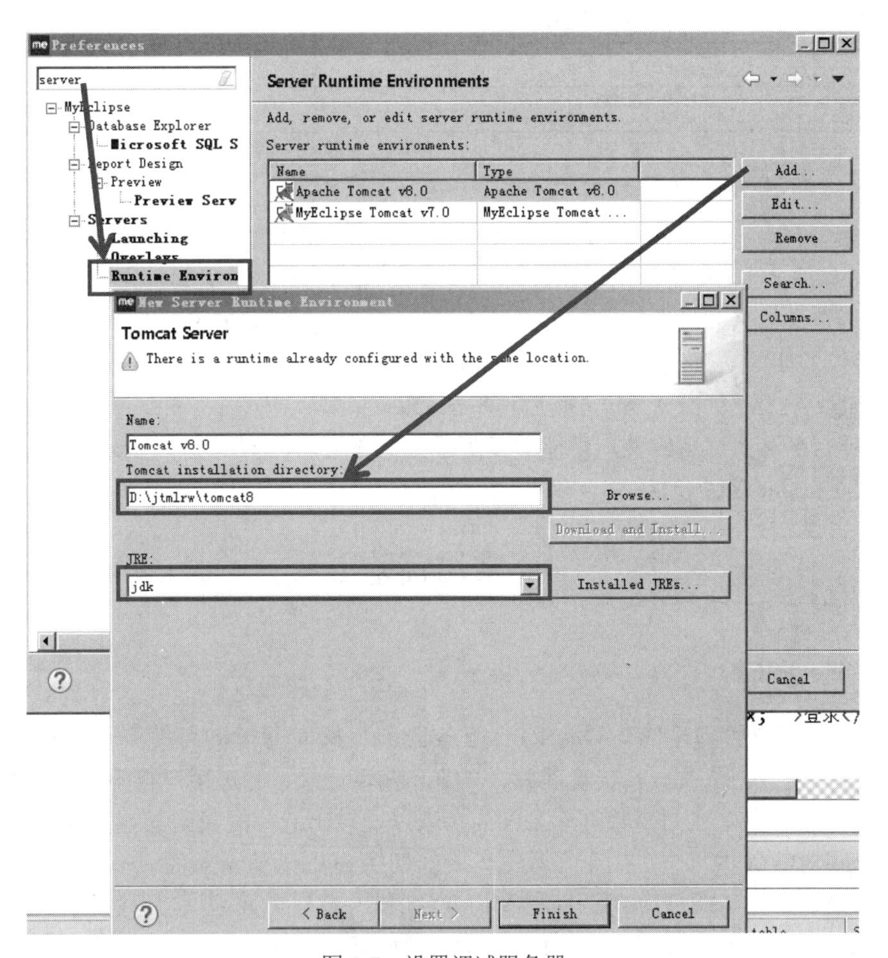

图 1-7　设置调试服务器

（7）选择 MyEclipse 中的 maven4myeclipse，为了避免长时间联网更新，只勾选 Offline、Do not automatically update，将 mavenjdk 改为 JTM 中的 jdk。如果要创建 Maven 项目，不能这样选择。

1.3.3 常用快捷键

为了提高编码、调试效率，需要掌握 MyEclipse 中一些常用快捷键的使用，见表 1-1。

表 1.1 MyEclipse 中常用的快捷键

快捷键	作用
Ctrl+Shift+F	格式化代码，规范选中的代码格式
Ctrl+Shift+O	在 Java 文件中使用，自动引入（import）需要的 Java 包或清理不需要的 Java 包
Ctrl+Shift+/	自动识别代码类型，为选中的内容添加相应的注释，如在 html 代码中添加 <!-- --> ，在 Java 或者 JavaScript 代码中添加/* */
Ctrl+Shift+\	取消相应的注释
Alt+/	代码提示、自动补全
Ctrl+M	窗口最大化和还原
Ctrl+D	删除一行
Ctrl+O	显示类中方法和属性的大纲

1.3.4 常见问题

（1）项目名称上出现红色感叹号，说明项目的 Jar 包引入的有问题（可能是工程中 java build path 中指向的 Jar 包路径错误，或者引入了非 Jar 包文件），需要重新引入 Jar 包。

（2）多个文件名前显示红叉，但文件来源于官方，内容应该没问题，这是 MyEclipse 校验报错。可以关闭 MyElipse 的校验。单击 Window→Preferences，选择 MyEclipse→Validation→选中 Disable ALL，单击 Apply，然后再单击 OK。

（3）项目名称上出现红叉，可能是复制了其他项目，但 JDK 或者 Runtime 配置不一致，导致默认引用的 Jar 路径异常。

1.4 常用辅助工具

1.4.1 HeidiSQL

HeidiSQL 是一款用于简化 MySQL 服务器和数据库管理的免费开源的图形化界面。HeidiSQL 软件允许浏览数据库、管理表、浏览和编辑记录、管理用户权限、从文本文件导入数据、运行 SQL 查询、在两个数据库之间同步表以及导出选择的表到其他数据库或者 SQL 脚本当中。HeidiSQL 提供了一个用于在数据库之间切换 SQL 查询和标签带有语法突出显示的简单易用的界面，其他功能包括 BLOB 和 MEMO 编辑、大型 SQL 脚本支持、用户进程管理等。

通过 HeidiSQL 工具管理 MySQL 数据库、创建表、编辑数据、设置字符集为 UTF8。

★非常重要★

1.4.2 Notepad++

Notepad++是程序员必备的文本编辑器，软件小巧高效，支持 27 种编程语言，如 C、C++、Java、C#、XML、HTML、PHP、JS 等，可完美地取代微软的记事本。配置 Notepad++到右键菜单，便于文档内容的编辑、批量查找替换。

1.4.3 MagicalTool

MagicalTool 是代码自动生成工具，支持模板，可以直接生成 DAO、Controller 和 Service 等；瞬间完成项目表的增删查改；把数据库的表或视图直接快速生成符合特定 Java 框架技术所需的 POJO 类。

1.5　考核任务

本章主要考核开发工具的准备、环境的配置、工具的安装。

开发环境检查考核明细如下：

（1）能正常打开 MyEclipse。（20 分）

（2）JDK 和 Tomcat 都在同一目录下，安装目录符合要求。（20 分）

（3）在命令窗口，输入 java –version 后按回车键，可以看到 Java 的版本信息。（20 分）

（4）运行 tomcat\bin\startup.bat 文件，在弹出的命令窗口中可以看到最后一行显示 Server startup in xxxxx ms。（20 分）

（5）运行 HeidiSQL，配置 IP（127.0.0.1）、用户名（root）、密码，可看到系统中的默认数据库，比如 MySQL。（20 分）

本章小结

本章主要介绍开发 JavaEE 项目需要的工具并进行下载与安装，对相关环境变量、工具参数的配置。只有当开发环境和工具都安装配置完善，才能进行 JavaEE 项目的快速开发，也就是“磨刀不误砍柴功”。

第 2 章　新闻发布系统设计

在开发 JavaEE 的 Web 应用系统时，首先要弄清楚用户的需求。简而言之，要系统做什么，必须了解和分析客户的需求，尽量准确地掌握客户需要一个什么样的软件系统。

在与客户充分地沟通协调后，得到一个双方都认可的需求分析报告，在此基础上，需要弄清楚怎么做，也就是如何设计并实现客户需要的软件系统。根据软件工程的概念，设计包括概要设计、详细设计，本书的重点是 JavaEE 框架技术的应用，所以只对新闻发布系统做了比较简单的设计，包括功能模块设计和数据库设计。

2.1　系统需求分析

本书以新闻发布系统（已经简化了新闻分类、图片新闻、置顶等功能）作为实战项目，让读者能够在较短的时间内用一种框架技术（Nutz、SSH、SSM 或 JFinal）实现系统的开发。

系统中有两种用户，普通用户和管理员，不同用户有不同的功能权限。

- 普通用户可以通过系统首页查看分页新闻列表，单击一条新闻后可以在新闻阅读页查看新闻详情。
- 管理员除了拥有普通用户的权限以外，还可以通过首页登录到后台管理页。在后台，可以发布新的新闻，可以查询新闻，修改指定的新闻，删除指定的新闻。

根据上述系统需求分析，设计系统功能模块如图 2-1 所示。

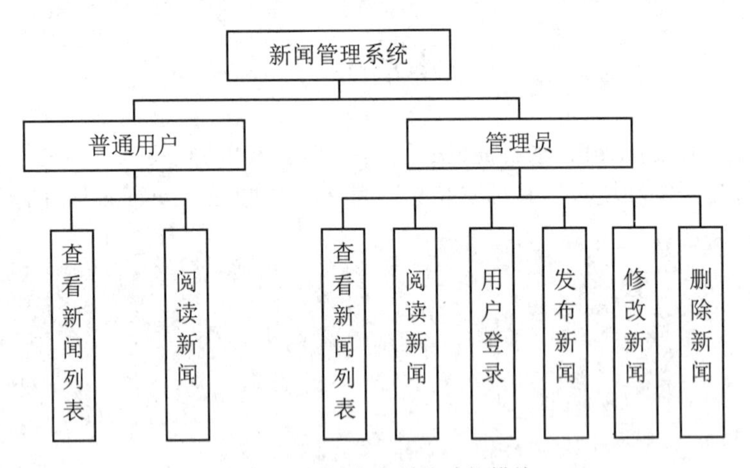

图 2-1　新闻发布系统功能模块

2.2　系统功能预览

不同的用户通过系统进行不同的操作，每个操作都是一个功能的体现，下面给出系统需要实现的具体功能。

2.2.1　查看新闻列表

普通用户和管理员都可以查看新闻列表。新闻列表以分页形式展示，每页显示 10 条新闻，每 5 条新闻后，显示一条分隔线，列表只显示新闻的标题、新闻发布日期，如图 2-2 所示。

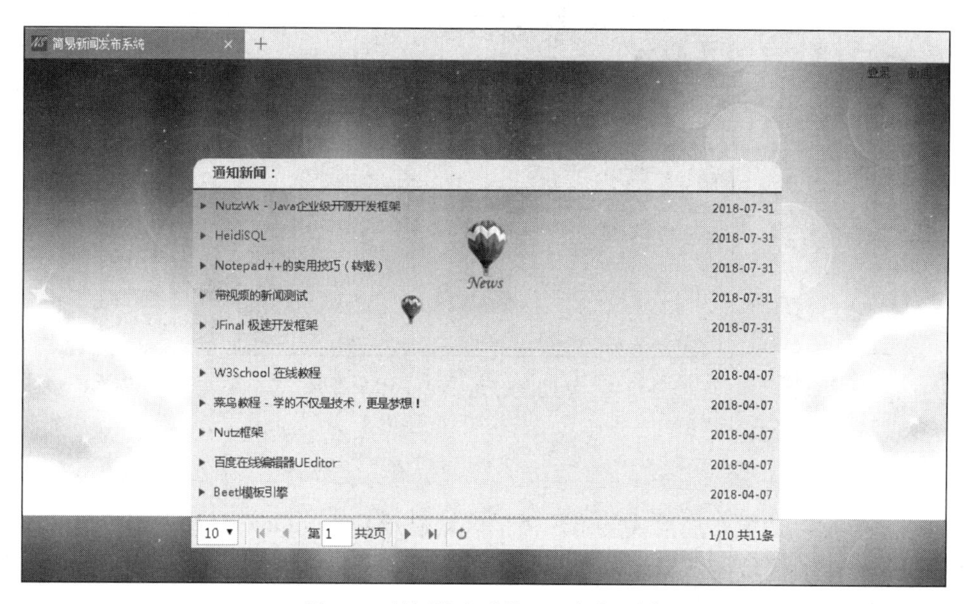

图 2-2　新闻发布系统——新闻列表

2.2.2　阅读新闻

单击新闻列表中任意一条新闻的标题，系统打开新闻阅读页面，页面显示新闻的标题、发布者、发布日期、阅读量、新闻内容，如图 2-3 所示。

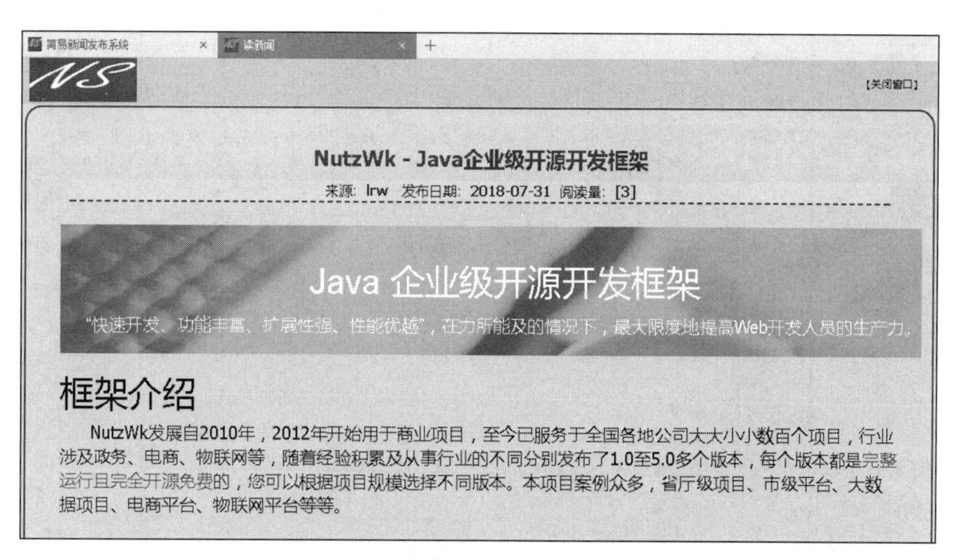

图 2-3　新闻发布系统——阅读新闻

2.2.3 用户登录

如果用户需要发布、修改或者删除新闻，则要以管理员身份登录。登录时输入用户名和密码，然后单击“登录”按钮或直接回车，如图 2-4 所示。

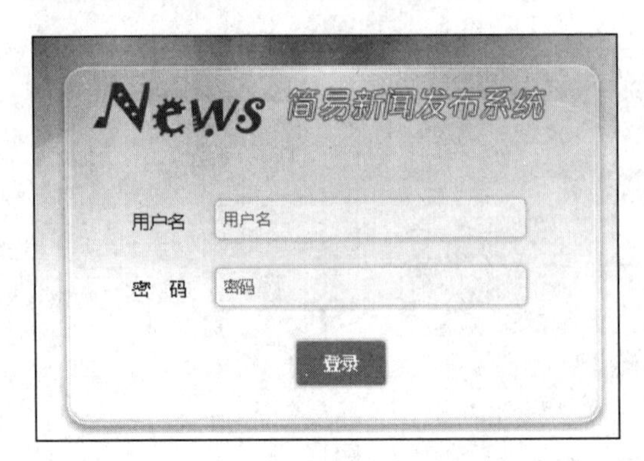

图 2-4 新闻发布系统——用户登录

2.2.4 发布新闻

后台管理界面以边框布局形式展示，顶部区域显示系统 LOGO、登录用户名；左部区域显示系统菜单；中心区域显示新闻发布区；底部区域显示版权信息。

发布新闻前，输入新闻标题、新闻发布者、新闻内容。新闻内容可以包含文字、图片、音频、视频、超链接等多媒体形式，可以设置字的大小、对齐方式、颜色、字型等。完成输入后，单击“保存”按钮，实现发布新闻功能，如图 2-5 所示。

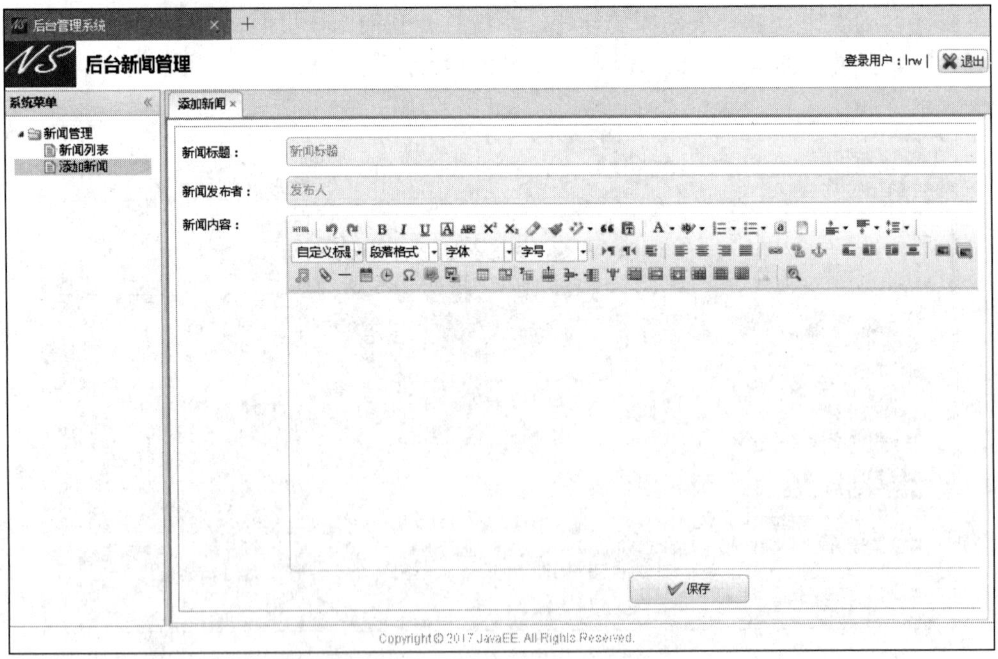

图 2-5 新闻发布系统——发布新闻

2.2.5　修改新闻

在新闻列表中选择待修改的新闻，修改新闻页面上将显示待修改新闻的标题、发布者、内容；不显示的项（如发布日期、阅读量等）不允许修改。完成修改后，单击“保存”按钮，实现修改新闻的功能，如图 2-6 所示。

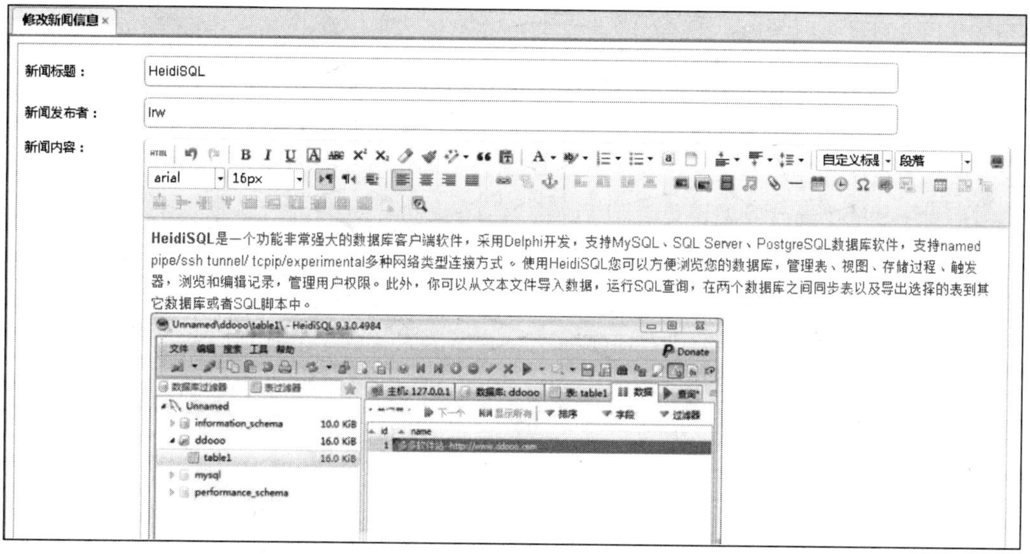

图 2-6　新闻发布系统——修改新闻

2.2.6　删除新闻

只能删除指定 ID 的新闻。为了防止误操作删除，在正式删除前，系统会弹出一对话框让用户确认，仅当用户单击“确定”按钮后才能删除一条新闻，如图 2-7 所示。

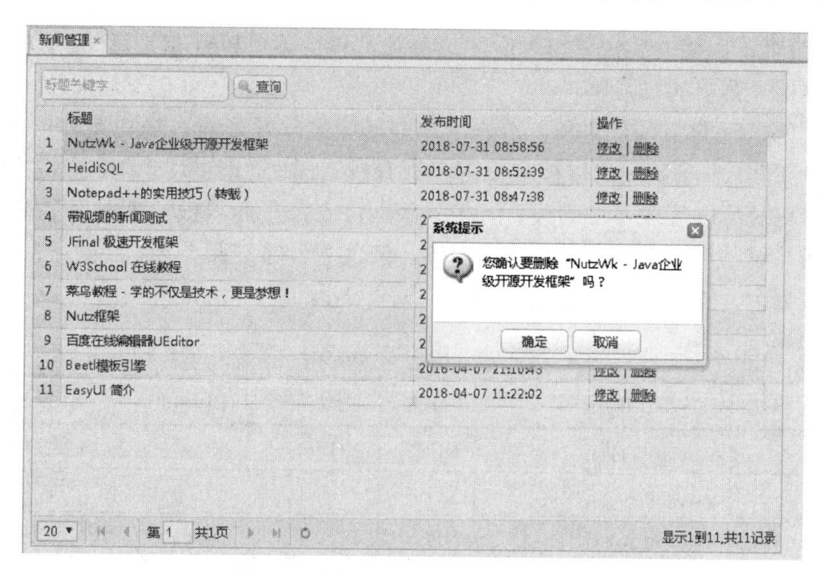

图 2-7　新闻发布系统——删除新闻

2.3 数据库设计

新闻发布系统应包含用户、新闻、系统菜单信息，该系统的实体联系（E-R）图如图 2-8 所示。

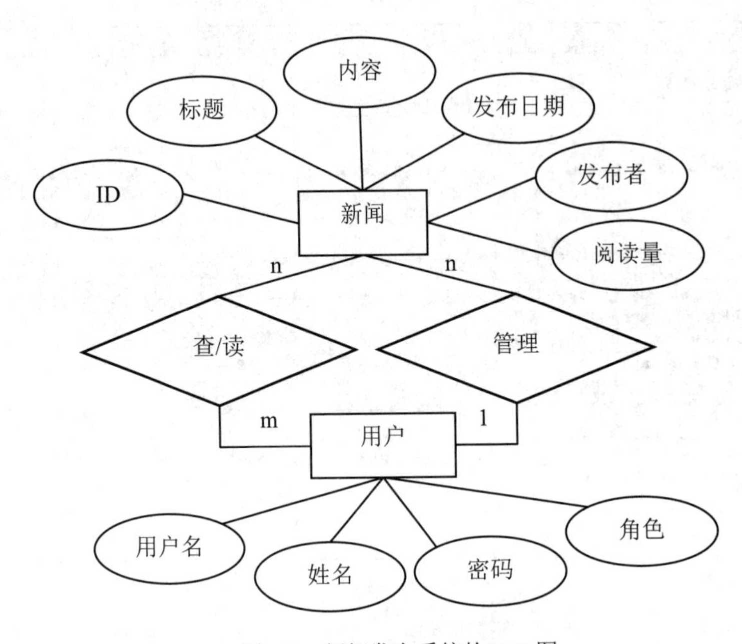

图 2-8 新闻发布系统的 E-R 图

（1）用户：代表系统的用户实体，主要包括用户的用户名、姓名、密码和角色。

（2）新闻：代表新闻信息，包括新闻的 ID、新闻标题、新闻内容、发布时间、发布者（人或机构）、阅读量（单击量）。

（3）系统菜单：代表后台管理页面的菜单项，包括菜单栏目 ID、父级 ID、菜单栏目名称、菜单的链接、菜单的权限用户。

用户与新闻之间的关系有两种，普通用户与新闻之间是多对多的关系，一个用户可以查看或阅读多条新闻，一条新闻可以被多个用户查看或阅读；管理员用户与新闻之间是一对多的关系，一个管理员可以管理（增删改查）多条新闻，一条新闻只被一个管理员管理。

新闻发布系统需要 3 张表，分别为用户表、新闻表和系统菜单表。

登录系统后台，需要用户信息表，表结构见表 2-1。

表 2-1 用户表（user）

Field	Type	Null	Key	Default	Comment
uid	varchar(11)	NO	PRI		用户名
xm	varchar(20)	NO			姓名
pwd	varchar(50)	NO			密码
role	char(1)	NO			角色 1 管理员

创建 user 表的 SQL 代码如下：

```
CREATE TABLE `user` (
`uid` VARCHAR(11) NOT NULL COMMENT '用户名',
`xm` VARCHAR(20) NOT NULL COMMENT '姓名',
`pwd` VARCHAR(50) NOT NULL COMMENT '密码',
`role` CHAR(1) NOT NULL COMMENT '角色',
PRIMARY KEY (`uid`)
)
COMMENT='用户信息'
COLLATE='utf8_general_ci'
ENGINE=InnoDB;
```

新闻表用来保存管理员发布的新闻，表结构见表 2-2。

表 2-2　新闻表（news）

Field	Type	Null	Key	Default	Comment
id	int(11)	NO	PRI	auto_increment	新闻 id
title	varchar(255)	NO			新闻标题
content	text	NO			新闻内容
tjdate	datetime	NO			发布时间
cruser	varchar(11)	NO			发布者
hitnum	int(10)	NO		0	阅读量

创建 news 表的 SQL 代码如下：

```
CREATE TABLE `news` (
`id` INT(11) NOT NULL AUTO_INCREMENT COMMENT '新闻 ID',
`title` VARCHAR(255) NOT NULL DEFAULT '0' COMMENT '新闻标题',
`content` TEXT NOT NULL COMMENT '新闻内容',
`tjdate` DATETIME NOT NULL COMMENT '发布时间',
`cruser` VARCHAR(11) NOT NULL COMMENT '发布者',
`hitnum` INT(10) NOT NULL DEFAULT '0' COMMENT '阅读量',
PRIMARY KEY (`id`)
)
COMMENT='新闻'
COLLATE='utf8_general_ci'
ENGINE=InnoDB;
```

管理员登录到后台时所见的菜单栏目保存在菜单表，表结构见表 2-3。

表 2-3　菜单表（cmenu）

Field	Type	Null	Key	Default	Comment
id	int(11)	NO	PRI	auto_increment	栏目 id
pid	int(11)	NO		0	父级 id
name	varchar(50)	NO			栏目名称
url	varchar(100)	YES			链接
permission	varchar(50)	YES			权限标识

创建 cmenu 表的 SQL 代码如下：

```
CREATE TABLE `cmenu` (
`id` INT(11) NOT NULL AUTO_INCREMENT COMMENT '栏目 ID',
`pid` INT(11) NOT NULL DEFAULT '0' COMMENT '父级 ID',
`name` VARCHAR(50) NOT NULL COMMENT '栏目名称',
`url` VARCHAR(100) NULL DEFAULT NULL COMMENT '链接',
`permission` VARCHAR(50) NOT NULL COMMENT '权限标识',
PRIMARY KEY (`id`)
)
COMMENT='系统菜单'
COLLATE='utf8_general_ci'
ENGINE=InnoDB;
```

2.4 考核任务

考核任务包括系统的需求分析、系统的功能模块、数据库设计：

（1）系统的需求分析。（30 分）

（2）系统的功能模块。（30 分）

（3）数据库设计。（40 分）

本章小结

在项目实战开始前，本章对系统的需求分析、系统设计、系统功能进行了介绍，使读者对开发系统有了较全面的了解；另外对设计新闻发布系统需要的数据库表进行了设计和创建。

第3章　基于Nutz的项目实战

现在项目采用快速迭代的开发方法，边设计边开发，开发完一个功能就测试一个功能，循环渐进。每一个完成的功能块，具备可视、可集成和可运行的特点。这种能及时看到运行效果的开发方法，可以帮助学习者快速获得成就感、培养开发兴趣、树立信心。

3.1　Nutz框架简介

Nutz 是一组轻便小型的框架集合，各个部分可以被独立使用。Nutz 的目标是在不损耗运行效率的前提下，让 JavaEE Web 项目的开发人员写更少的代码，因此可以获得更快的开发速度。

Nutz 框架有以下特点。

（1）遵循 Apache 协议，完整开源，永久免费。

（2）体积小巧（1M+）且无依赖。

（3）学习曲线极低，在线的使用手册、API 文档齐全，在 QQ 群、Nutz 社区提问，可很快得到回复。

（4）框架集成的功能强大，如 Nutz.Dao 比 Hibernate 控制和 iBatis 的 SQL 简便；Nutz.Ioc 配置比 Spring 配置简单；Nutz.Mvc 具有高可控的路由功能，即将 HTTP 的请求路由到用户的自定义函数，这个过程是高度可定制化的，同时内置了很多常用的路由方式；Nutz.Json 提供轻量级的 JSON 数据转换等优越功能；内置支持表单文件上传以及下载的断点续传；提供各种帮助函数。

（5）Nutz 的测试用例覆盖率很高，保证了框架本身的质量。

（6）应用广泛，适合服务器、Android、嵌入式等各种开发场景。

3.2　创建 Nutz 项目

3.2.1　项目工程结构

用 JavaEE 框架技术创建 Web 项目前，最好先规划好项目工程结构，分类科学管理各种资源。例如采用表 3-1 所示的项目工程结构，不同的文件夹或包（package）中存放不同类型的文件资源。详细情况可以参照表中的说明。

在 Java 源代码文件夹 src 下的是包（package），package 的命名通常符合下述规则：

（1）包名由小写字母和圆点组成

（2）包的路径，一般形如 com.公司名/团队名/个人名.项目名.分类名……，基本做到见名知义。比如 cn.lrw.newsnutz.pojo 中，cn 代表中国，lrw 代表个人，newsnutz 代表项目名称，pojo 表示当前包里面存放数据库表/视图对应的 Java 类。

表 3-1 项目工程结构

路 径	说 明
newsnutz	项目名称
├─**src**	源码文件夹
│ ├─cn.lrw.newsnutz	Main 包，主模块，存放 MainModule，MainSetup 类
│ ├─cn.lrw.newsnutz.pojo	POJO 包，子模块，存放数据库表/视图对应的 Java 类
│ ├─cn.lrw.newsnutz.module	Module 包，子模块，存放业务逻辑类
│ └─cn.lrw.newsnutz.utils	Utils 包，子模块，存放封装公用方法或属性的类
├─**conf**	存放 JS、JSON、Properties 等类型的配置文件
│ ├─ioc	存放诸如数据源的配置文件
│ └─dao.js	数据库配置文件
│ ├─log4j.properties	系统日志配置文件
│ ├─config.json	UEditor 的配置文件
│ └─beetl.properties	Beetl 模板配置文件
└─**WebRoot**	Web 项目根目录
├─error	存放异常访问提示页，如 403. html、404. html 等
├─include	分类存放网页中引用的 JS、CSS、图像类文件资源
│ ├─css	
│ ├─js	
│ ├─img	
│ ├─easyui	EasyUI 框架
│ └─ueditor	百度提供的可视化 HTML 编辑器
├─upload	存放上传的文件
├─WEB-INF	Java 的 Web 应用的安全目录，自动存放 class 文件
│ ├─lib	存放项目需要的 Jar 包
│ ├─web	存放网站中的网页文件，如 htm、html、jsp 等
│ └─web.xml	Web 工程的配置文件
└─index. html	Web 工程默认的首页文件

3.2.2 准备 Jar 包和 JS 库

现在项目主要使用表 3-2 所示的 Jar 包，可以从各自的官方网站下载最新版本，也可以从 Maven 中央仓库查找和下载。Maven 中央仓库包含了世界上大部分流行的开源项目构件的 Jar 包，有很多 Jar 包还提供各种历史版本。

当前项目还使用到一些常用的 JS（JavaScript）库、CSS 库文件，如 jQuery、EasyUI、UEditor 等，它们都可以从官方网站下载到本地计算机上，也可以通过 BootCDN 网站提供的下载地址下载。

表 3-2　项目所需的主要 Jar 包

Jar 包的名称	说　明
Nutz	一组轻便小型的框架集合
mysql-connector-java	MySQL 数据库驱动
druid	数据库连接池 Druid，带强大的监控功能
Log4j	用 Java 编写的可靠，快速和灵活的日志框架
beetl	一款超高性能的 Java 模板引擎
shiro	用于认证、授权、加密、会话管理、与 Web 集成、缓存等
ueditor	百度在线编辑器。依赖 json.jar、commons-codec-1.10.jar、commons-io-2.4.jar、commons-fileupload-1.3.3.jar、commons-lang3-3.2.1.jar

建议建一个专门的文件夹，把下载的 Jar 包和 JS 库、CSS 库等资源集中存放，便于以后其他项目的开发。

3.2.3　新建 Web 项目

运行 MyEclipse2015，选择 File 菜单中的 New 命令，单击 Web Project，打开 New Web Project 窗口，如图 3-1 所示，设置基本参数如下：

- Project name：输入适当的项目名称，例如 newsnutz，建议项目名称由字母或数字组成，名称短小但见名知义。
- Java EE version：选择 MyEclipse2015 中最新的 JavaEE 7，对应 Web 3.1 的版本。
- Java version：选择本机安装的 1.8 版本。
- JSTL Version：由于当前项目不使用 jsp 页面，所以 JSTL Version 选择 none，也就是不需要 JSTL 支持。
- Target runtime：选择运行环境为本机安装的 Tomcat v8.0。

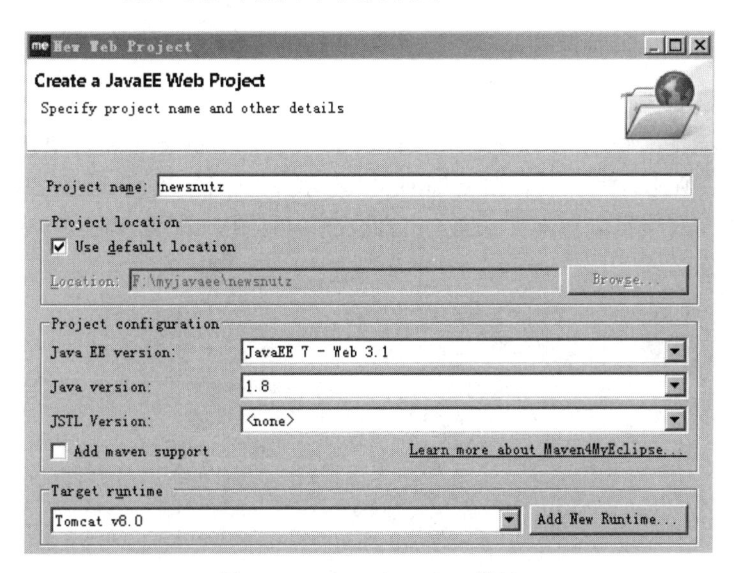

图 3-1　New Web Project 窗口

设置完成上述参数后，单击 Next 按钮，进入源文件夹（Source Folder）配置窗口，如图 3-2 所示。在当前窗口，可以看到 src 中的 java 文件在经过编译后，默认输出文件夹（Default Output Folder）为 WebRoot\WEB-INF\classes。根据项目工程结构，conf 文件夹中的配置文件也应该输出到 WebRoot\WEB-INF\classes 目录，所以在这个窗口中，单击 Add Folder 添加文件夹 conf。

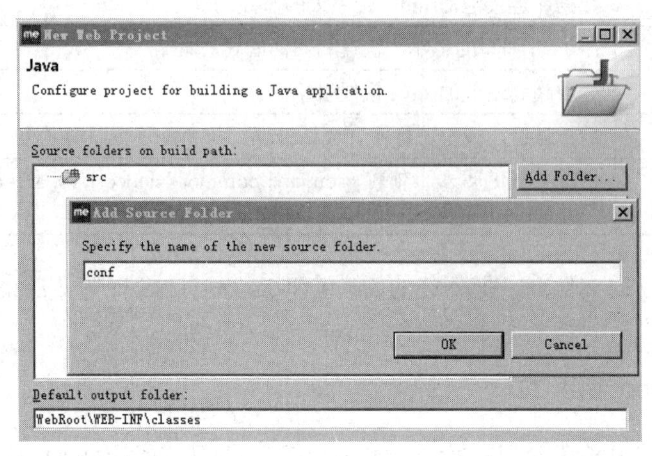

图 3-2　Source folders on build path 窗口

存放配置文件的 conf 文件夹添加完成后，单击 Next 按钮进入模块配置（Web Module）窗口，如图 3-3 所示。

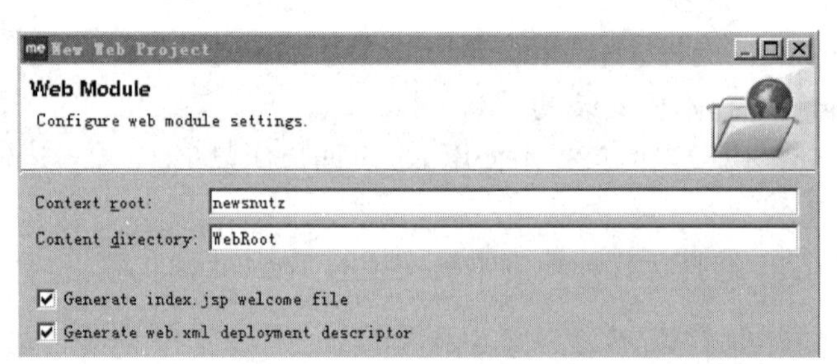

图 3-3　Web Module 窗口

在模块配置窗口，勾选 Generate web.xml deployment descriptor，单击 Finish，生成以 newsnutz 为名称的 Web 项目，项目结构如图 3-4 所示，其中在 WebRoot\WEB-INF 目录下可以看到自动生成的 Web 项目配置文件 web.xml。此文件主要用于配置欢迎页 welcome-file、Filter、Listener、Servlet 等。

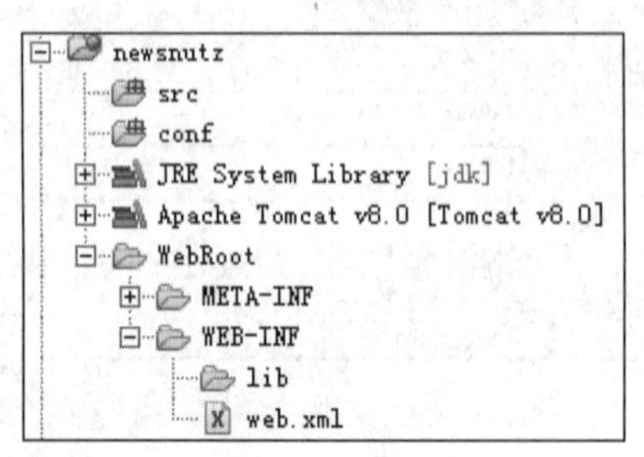

图 3-4　项目结构

将准备好的 Jar 包添加到 WebRoot/WEB-INF/lib 目录下，效果如图 3-5 所示。百度编辑器的压缩包中包含了 commons-fileupload-1.3.1.jar，百度官方声称 commons-fileupload-1.3.1.jar 存在漏洞，可能会导致 DDOS 攻击，强烈推荐升级至最新版本 commons-fileupload-1.3.3.jar。

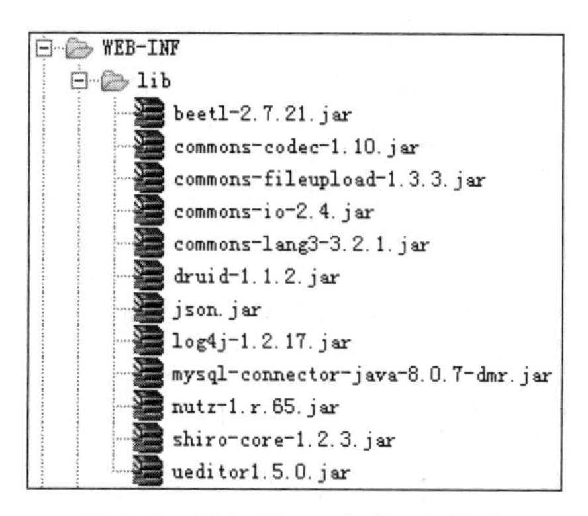

图 3-5 添加了 Jar 包的 lib 目录

参照“项目工程结构”添加包（package）和其他文件夹，完成效果如图 3-6 所示。

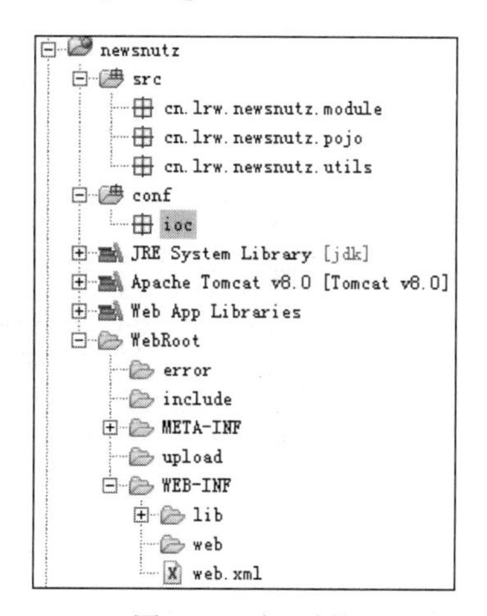

图 3-6 项目结构

3.2.4 添加数据源

1. 创建数据库

利用可视化工具 HeidiSQL 创建数据库 dbnews1。创建数据库时应注意，要保证数据库的字符集是 UTF-8，选择 Collation 的值为 utf8_general_ci，如图 3-7 所示。

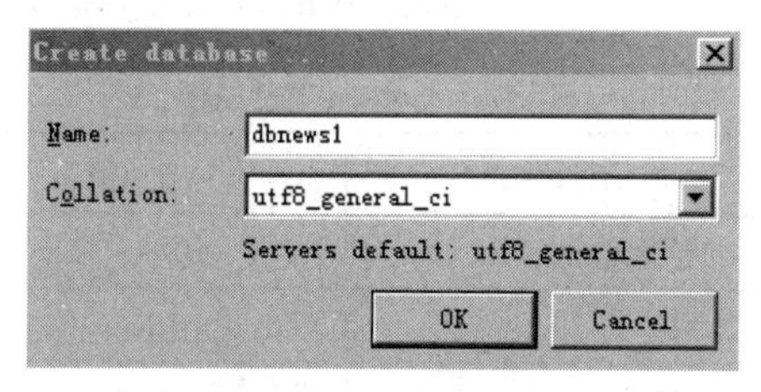

图 3-7 新建数据库

软件国际化是大势所趋，Unicode 是国际化最佳的选择。MySQL 有两个支持 Unicode 的 character set：UCS2 和 UTF-8，一般选择 UTF-8。

每个 character set 会对应一定数量的 collation，collation 即比对方法，用于指定数据集如何排序，以及字符串的比对规则。

collation 名字的规则可以归纳为如下两类：

- <character set>_<language/other>_<ci/cs>
- <character set>_bin

例如：

- utf8_ general _ci/cs：ci 是 case insensitive 的缩写；cs 是 case sensitive 的缩写。它们指定大小写是否敏感。
- utf8_bin：表示将字符串中的每一个字符用二进制数据存储，区分大小写。

选择了 Collation，就确定了字符集。例如选择 utf8_ general _ci，就确定了字符集是 UTF-8。

2. 创建表并录入数据

创建数据库以后，根据第 2 章中表 2-1 所示的表结构，在 dbnews1 数据库中创建 user 表，如图 3-8 所示。

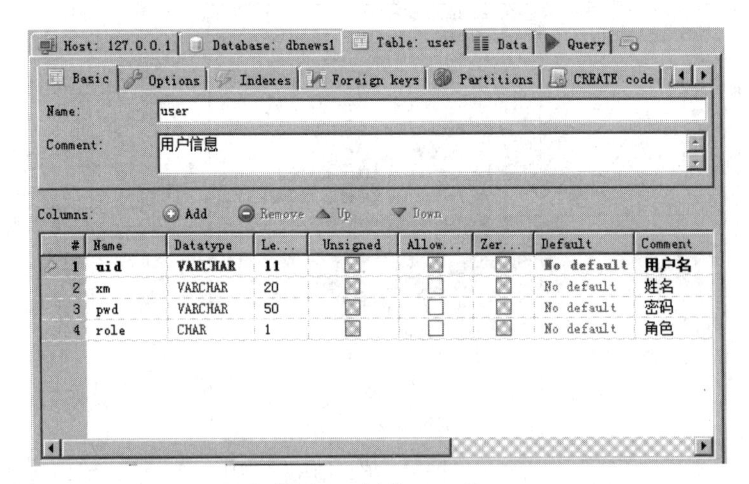

图 3-8　创建 user 表

根据第 2 章中表 2-2 所示的 news 表结构和表 2-3 所示的 cmenu 表结构，用类似的方法在 dbnews1 数据库中创建 news 表和 cmenu 表。

cmenu 中数据的组织是按 id 和 pid 标识层次关系，name 值为显示在界面上的菜单名称，url 为请求打开指定功能的页面的路径，permission 设定为指定角色的用户才能看到相应的菜单项。通过 HeidiSQL 等可视化工具直接将 cmenu 表的数据录入数据库。cmenu 表数据见表 3-3。

表 3-3　cmenu 表数据

id	pid	name	url	permission
1	0	新闻管理		1
2	1	新闻列表	/news/goList	1
3	1	添加新闻	/news/goAdd	1

3. 创建配置文件

在 conf/ioc 目录创建 dao.js 文件，该文件的主要作用是创建 dataSource，配置项及说明如图 3-9 所示。这是 Nutz 项目中最常见的配置方式，由 NutIoc 来管理 dataSource 和 NutDao 的实例 DAO。

```
var ioc = {
    dataSource : { //定义数据源对象
        //采用阿里巴巴出品的druid连接池实现dataSource
        type : "com.alibaba.druid.pool.DruidDataSource",
        events : {//声明对象的事件。
            create : "init",//当这个对象被创建后，会调用这个接口的唯一方法
            depose : 'close'// 当对象被容器销毁时触发。
        },
        //配置连接池中的参数
        fields : {
            url : "jdbc:mysql://127.0.0.1:3066/dbnews1?serverTimezone=Hongkong&useSSL=false",
            username : "root",
            password : "",
            testWhileIdle : true, // 非常重要,预防mysql的8小时timeout问题
            maxActive : 100,//连接池的最大数据库连接数。
            maxWait : 5000 //最大建立连接等待时间。在5秒内拿不到新连接,就抛异常,避免druid一直尝试获取连接,会导致假死
        }
    },
    dao : {
        type : "org.nutz.dao.impl.NutDao",//对象dao的实现类，new NutDao()
        args : [{refer:"dataSource"}] // 引用数据源作为构造函数的参数
    }
};
```

图 3-9　创建 dataSource

JavaEE 的很多项目中会用到控制反转 IoC。IoC 是 Inversion of Control 的缩写，是面向对象编程中的一种设计思想，通过将对象的控制权转移给 IoC 容器，不直接在类内部通过 new 创建对象，而是由 IoC 容器查找及注入依赖对象，类只是被动地接受依赖对象，从而降低代码之间的耦合度。

Nutz.Ioc 是将一部分关于对象的依赖关系单独存储在某种介质里，并且提供一个接口帮助使用者获得这些对象。Nutz.Ioc 核心逻辑并没有限定配置信息的存储方式，但它还是提供了一个默认的 JSON 或者 JS 配置文件。选择 JSON 格式，有两个优点：

- 省却了 XML 书写的烦恼。
- 避免了硬编码。当修改了配置时，不需要重新编译工程。

在配置过程中，根据项目数据库的实际情况，正确设置数据库的驱动、地址、端口、数据库名、用户名和密码。

在设定时区的时候，如果设定 serverTimezone=UTC，会比中国时间早 8 个小时。所以在中国，可以设置 serverTimezone=Hongkong

高版本 MySQL（版本号高于 5.7），需要指明是否进行 SSL 连接，如果不需要 SSL 连接，则设定参数 useSSL=false，否则会在 Console 显示如下信息：

WARN: Establishing SSL connection without server's identity verification is not recommended。

3.2.5　DAO 注解

Java 类中会经常见到注解（Annotation），它会使得编程更加简洁、代码更加清晰。Java 注解是从 JDK1.5 引入，目前很多主流框架都支持注解，编写代码时应尽量使用注解。

1. 注解的定义

注解是一种代码级别的说明，与类、接口、枚举在同一个层次。它可以声明在包、类、字段、方法、局部变量、方法参数等的前面，用来对这些元素进行说明和注释。

2. 注解的分类

（1）按照运行机制划分。

- 源码注解：只在源码中存在，编译成.class 文件注解就不存在了。
- 编译时注解：在源码和.class 文件中都存在，如，@Override、@Deprecated、@SuppressWarnings 都属于编译时注解。
- 运行时注解：在运行阶段起作用，如，@Autowired 自动注入，在程序运行时把成员变量自动地注入文件。

（2）按照来源划分。

- JDK 注解：主要有 3 个，@Deprecated、@Override、@SuppressWarnings。
- 第三方注解：如，@Autowired、@Inject、@Table。
- 自定义注解：根据需要自己定义的一些注解。
- 元注解：给注解进行注解，可以理解为注解的注解，如，@Retention、@Documented、@Target、@Inherited、@Repeatable。

（3）按照参数的个数划分。

- 标记注解：没有参数。
- 单值注解：如果单值的名称是 value，则在使用时可以省略“value=”，直接写单值。
- 完整注解：定义注解时，一般会包含一些可以设置默认值的元素。

注解的语法比较简单，除了@符号的使用之外，基本与Java固有语法一致。比如@Override，表示当前定义的方法将覆盖超类或接口中的方法。

主流框架已经定义了很多便于使用的注解，我们更多地是关注它们的使用。使用注解最主要的部分在于对注解的处理（注解处理器）。注解处理器就是通过反射机制获取被检查类或方法上的注解信息，然后根据注解元素的值进行特定的处理。

Nutz 框架中 DAO 支持的注解见表 3-4。

表 3-4　Nutz.Dao 支持的全部注解

注　解	说　明
@Column	字段
@ColDefine	字段精确定义
@Default	默认值
@EL	字段表达式宏
@Id	数值主键
@Name	字符主键
@PK	复合主键
@Many	一对多映射
@ManyMany	多对多映射
@One	一对一映射

续表

注　解	说　明
@Prev	自动设置
@Next	自动获取
@Readonly	只读声明
@SQL	字段 SQL 宏
@Table	表名
@View	视图名
@TableMeta	表设置
@TableIndexes	表索引
@Index	具体的索引内容
@Comment	表或者字段的注释

3.2.6　添加 POJO 类

POJO（Plain Ordinary Java Object）是指简单的 Java 对象，实际就是普通 JavaBeans，仅包含属性及其 getter/setter 方法的类，没有业务逻辑，可以理解为简单的实体类，用于封装数据库中的数据表。POJO 类对象可以方便地调用其 getter/setter 方法。

在 cn.lrw.newsnutz.pojo 包中创建 user 数据库表对应的 POJO 类 User，然后添加表中字段对应的属性和 getter/setter 方法。可以借助第三方工具，如 MagicalTools，批量自动生成数据库中大量表/视图对应的 POJO 类，这样可以大幅度减少编写代码的时间，提高开发效率。

可以通过“附录　在线资源”中提供的网址，下载 MagicalTools 工具。如果已经下载，运行 MagicalTools 目录中的 run.bat 程序，将会打开一个可视化窗口，如图 3-10 所示。在窗口中进行以下操作，可批量快速生成 POJO 类。

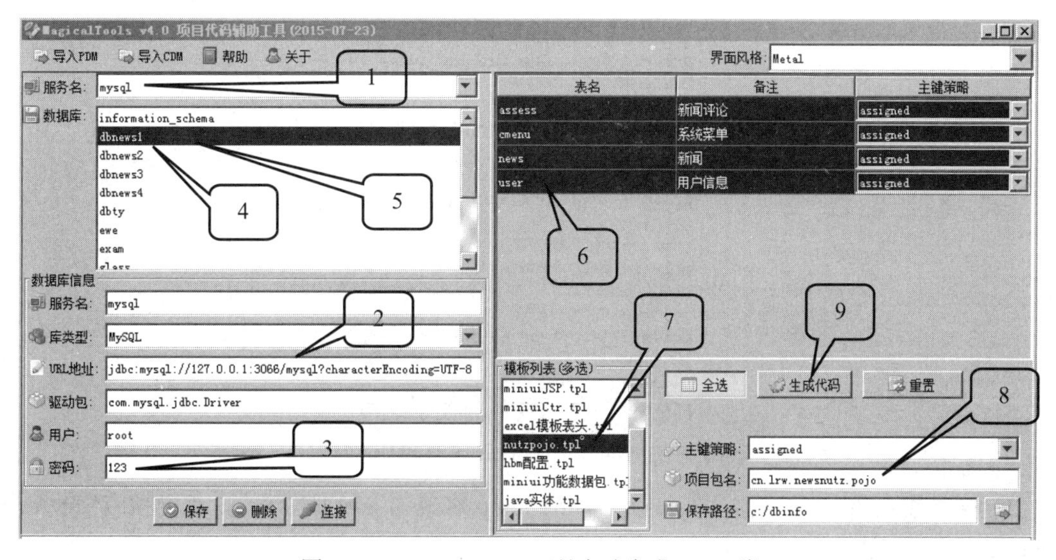

图 3-10　MagicalTools 工具自动生成 POJO 类

操作步骤：

（1）选择服务名 mysql。

（2）根据实际情况修改 URL 地址中的端口号、数据库和编码，如 3066，已经存在的数据库 mysql。

（3）输入连接 MySQL 数据库系统的密码，然后单击“连接”按钮。

（4）连接成功后，会在数据库列表中显示 MySQL 中的数据库名称，如 dbnews1。

（5）选择 dbnews1，就会显示 dbnews1 中的所有表和视图，如 user、news、cmenu。

（6）选择需要生成 pojo 的表和视图，如 user，也可以选择全部。

（7）选择模板列表中的模板“java 实体.tpl”，也可以参照已有的模板或自定义新的模板，如 nutzpojo.tpl。

（8）输入项目包名，如 cn.lrw.newsnutz.pojo。

（9）单击“生成代码”按钮将在默认的保存路径 C:\dbinfo\entities 下生成 POJO 类文件，如 User.java。

（10）把生成的 POJO 类文件复制到 MyEclipse 的 newsnutz 项目的 cn.lrw.newsnutz.pojo 包中。

（11）打开 POJO 类文件，注意 POJO 类中使用的注解，如@Table，@Column，@Id，@Name。

（12）数据库表中的字段对应到 POJO 类中的属性，属性前的注解通常是@Column，标记为字段。

（13）修改、生成 POJO 类文件。

1）“主键”属性前的注解应该为@Id（数值型主键）或@Name（字符型主键），如图 3-11 所示，新闻的 ID 是整数类型，所以注解修改为@Id。

```
@Table("news")
public class News  {

    @Id
    @Comment("新闻ID")
   private java.lang.Integer id;//
    @Column
    @Comment("新闻标题")
   private java.lang.String title;//
    @Column
    @Comment("新闻内容")
   private java.lang.String content;//
    @Column
    @Comment("发布时间")
   private Date tjdate;//
    @Column
    @Comment("发布者")
   private java.lang.String cruser;//
    @Column
    @Comment("阅读量")
   private java.lang.Integer hitnum;//
```

图 3-11　MagicalTools 生成的 POJO 类

2）将属性 tjdate 的日期类型 java.sql.Timestamp 修改为 java.util.Date，还要相应修改 tjdate 的 getter/setter 方法。

修改之后，如果文件中报错，使用快捷键 Ctrl+Shift+O 自动整理 import 引入的包，消除报错。修改完成后，保存文件。

补充：在 Web 开发中，避免不了对日期的操作，要注意表 3-5 所示的 4 种常见日期类的区别。

表 3-5　Java 日期类的区别

日期类	日期格式
java.util.Date	年月日时分秒
java.sql.Date	年月日（只存储日期数据不存储时间数据）
java.sql.Time	时分秒
java.sql.Timestamp	年月日时分秒纳秒（毫微秒）

3.2.7　创建主模块类

作为项目启动时的入口，Nutz 项目的主模块类类似于 Java 应用程序的主函数 main，是程序的入口函数。

任何一个类都可以作为主模块，只要在 web.xml 内正确配置即可。通常将主模块类命名为 MainModule，见名知义。在 cn.lrw.newsnutz 包中创建名称（Name）为 MainModule 的主模块类（Class），如图 3-12 所示。

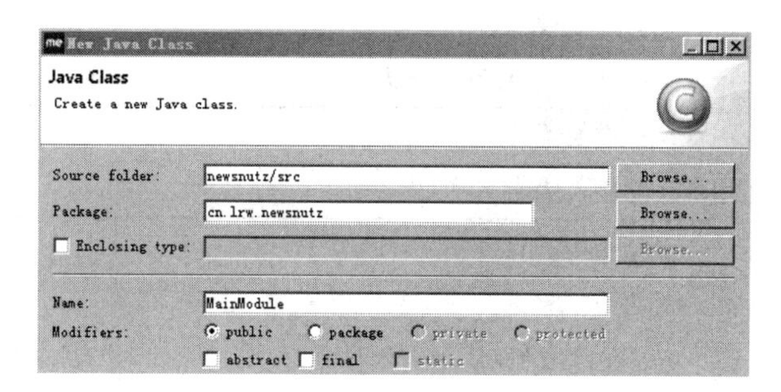

图 3-12　创建主模块类 MainModule

MainModule 类生成后，在类的前面添加注解，如图 3-13 所示。

```
@SetupBy(value=MainSetup.class)
@Modules(scanPackage=true)
@IocBy(type=ComboIocProvider.class, args={"*js", "ioc/","*anno", "cn.lrw.newsnutz","*tx"})
@Views({BeetlViewMaker.class})
public class MainModule {

}
```

图 3-13　主模块类 MainModule 及注解

当前项目主模块类 MainModule 前面使用的注解：

（1）@SetupBy(value=MainSetup.class)，需要新建一个类，类名可以是 MainSetup，实现 Setup 接口。在 MainSetup 类中，实现整个项目系统启动或者关闭时的一些处理工作，如初始化数据库、关闭 MySQL 线程等。

（2）@Modules(scanPackage=true)，将自动搜索主模块所在的包（包括子包）下所有的类。

（3）@IocBy(type=ComboIocProvider.class, args={"*js", "ioc/","*anno", "cn.lrw.newsnutz", "*tx" }），设置应用所采用的 Ioc 容器。

- ComboIocProvider 的 args 参数，星号开头的是类名或内置缩写，剩余的是各加载器的参数。
- *js 是 JsonIocLoader，负责加载 js/json 结尾的 IoC 配置文件。
- *anno 是 AnnotationIocLoader，负责处理注解式 IoC，例如@IocBean。
- *tx 是 TransIocLoader，负责加载内置的事务拦截器定义。

（4）@Views({BeetlViewMaker.class})，指定采用 Beetl 模板渲染视图。

能够用在主模块上的注解见表 3-6。

表 3-6 主模块上支持的注解

注 解	说 明
@Modules	整个应用有哪些子模块，子模块不能再嵌套子模块
@IocBy	整个应用应采用何种方式进行反转注入。如果没有声明，整个应用将不支持 IoC
@Localization	整个应用的本地化字符串的目录
@SetupBy	应用启动和关闭时，应该进行的处理
@Views	扩展整个应用支持的视图模板类型
@Ok	整个应用默认的成功视图
@Fail	整个应用默认的失败视图
@AdaptBy	整个应用默认的 HTTP 参数适配方式
@Filters	整个应用默认的过滤器数组
@Encoding	整个应用默认的输入输出字符编码

3.2.8 实现 Setup 接口

在整个应用启动或者关闭时，通常可以让系统做一些额外的处理工作，如读取配置文件、初始化数据库等。在 cn.lrw.newsnutz 包中新建 MainSetup 类，实现 Setup 接口，如图 3-14 所示。

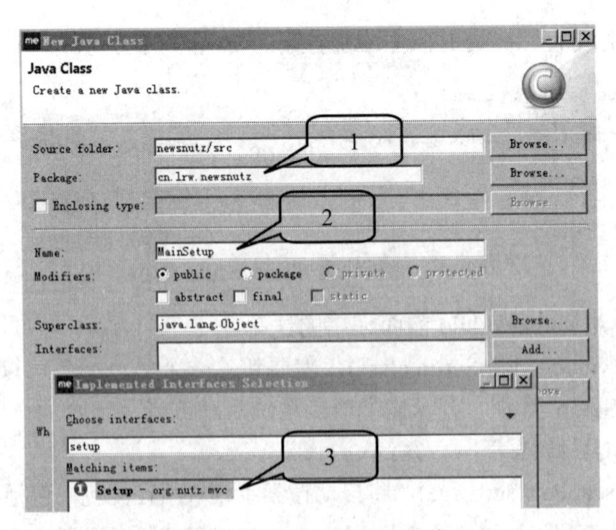

图 3-14 创建 MainSetup 类

在 MainSetup 类中初始化数据库，如果 user 表没有用户数据，则添加一条默认的用户信息：

- 用户名 2953，实际为登录用户名。
- 姓名 lrw。
- 密码 123，后期再运用 shiro 进行加密处理，提升安全性能。

```
public class MainSetup implements Setup {
	@Override
	public void init(NutConfig arg0) {
		Ioc ioc = arg0.getIoc();
		Dao dao = ioc.get(Dao.class);
		Daos.createTablesInPackage(dao, "cn.lrw.newsnutz.pojo", false);
		// 初始化默认根用户
		if (dao.count(User.class) == 0) {
			User user = new User();
			user.setUid("2953");
			user.setXm("lrw");
			user.setPwd("123");
			user.setRole("1");
			dao.insert(user);
		}
}
@Override
	public void destroy(NutConfig arg0) {// webapp 销毁之前执行的逻辑
	}
}
```

3.2.9　配置 web.xml

打开 web.xml 文件，在 display-name 节点和 welcome-file-list 节点之间添加 Nutz 的 Filter，根据实际情况正确设置 NutFilter 的 modules 参数值为主模块的包 cn.lrw.newsnutz.MainModule。

```
<display-name>newsnutz</display-name>
<filter>
	<filter-name>nutz</filter-name>
	<filter-class>org.nutz.mvc.NutFilter</filter-class>
	<init-param>
		<param-name>modules</param-name>
		<param-value>cn.lrw.newsnutz.MainModule</param-value>
	</init-param>
	<init-param>
	<param-name>ignore</param-name>
	<param-value>^(.+[.])(jsp|png|gif|jpg|js|css|jspx|jpeg|html)$</param-value>
	</init-param>
</filter>
<filter-mapping>
	<filter-name>nutz</filter-name>
	<url-pattern>/*</url-pattern>
</filter-mapping>
<welcome-file-list>
<welcome-file>index.html</welcome-file>
</welcome-file-list>
```

3.2.10 简单的系统首页

在 WebRoot 目录单击右键，选择新建 HTML，创建一个名为 index.html 的首页，如图 3-15 所示。页面内容自定，可以默认，如果原来有 index.jsp，可以删除。

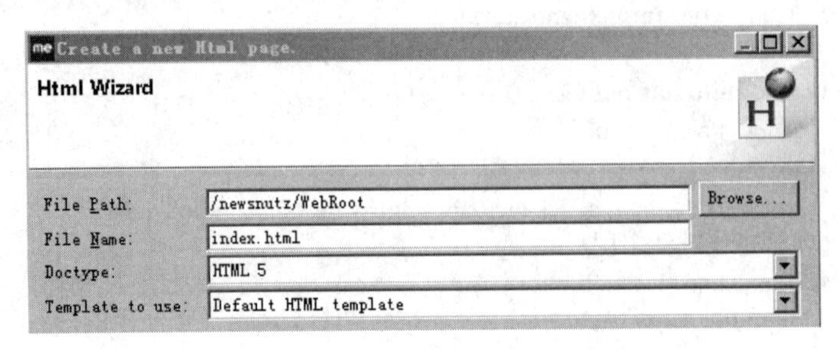

图 3-15 新建首页 index.html

3.2.11 运行项目

运行基于 Nutz 框架的 JavaEE 项目。如果控制台 Console 没有报错信息（出现 Exception 字样），则在浏览器可以看到正常的首页信息，数据库中可以看到写入的用户信息，如图 3-16 所示。

Host: 127.0.0.1 | Database: dbnews1 | Table: user | Data

dbnews1.user: 1 rows total (approximately)

uid	xm	pwd	role
2953	lrw	123	1

图 3-16 初始化数据库结果

首次运行，右键单击项目名称 newsnutz，选择 Run As 中的 Myeclipse Server Application 命令运行完成的项目，可以看到项目的运行状态，如图 3-17 和图 3-18 所示。

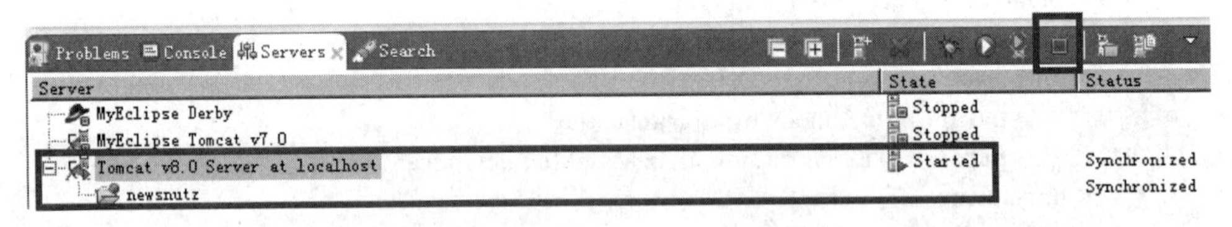

图 3-17 Myeclipse Server 运行状态

```
Problems Console Servers Search
九月23, 2018 11:46:47 上午 org.apache.catalina.startup.HostConfig deployDirectory
信息: Deploying web application directory D:\jtmlrw\tomcat8\webapps\ROOT
九月23, 2018 11:46:47 上午 org.apache.catalina.startup.HostConfig deployDirectory
信息: Deployment of web application directory D:\jtmlrw\tomcat8\webapps\ROOT has
九月23, 2018 11:46:47 上午 org.apache.coyote.AbstractProtocol start
信息: Starting ProtocolHandler ["http-apr-8080"]
九月23, 2018 11:46:47 上午 org.apache.coyote.AbstractProtocol start
信息: Starting ProtocolHandler ["ajp-apr-8009"]
九月23, 2018 11:46:47 上午 org.apache.catalina.startup.Catalina start
信息: Server startup in 2429 ms
```

图 3-18 控制台 Console 信息

在浏览器中输入http://localhost:8080/newsnutz并回车，如果能在浏览器中看到自己设定的首页，则说明系统运行正常，如图 3-19 所示。

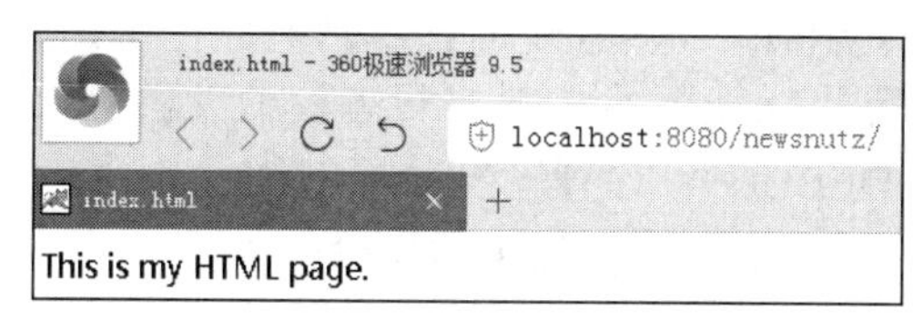

图 3-19　简单的默认首页

项目运行中，应学会查看 Console 中的信息，确认是否存在报错信息，排除存在的问题，比如：

- MainModule 参数出错。
- 数据库配置信息有误。
- jar 包没有添加到正确的目录中。
- web.xml 中参数不正确。
- server.xml 配置错误。
- MainSetup 中提前使用了加密方法。
- 启动 Tomcat 时，提示“Can’t load AMD 64-bit .dll on a IA 32-bit platform”，表示目前使用 64 位 Tomcat，32 位 JDK，重新安装 64 位 JDK 即可。

3.3　考核任务

掌握如何正确实现基于 Nutz 框架的入门项目的配置和开发过程。检查点如下：

（1）可以在浏览器看到默认首页 index.html 的内容，无乱码。(40 分)

（2）MyEclipse 的 Console 中没有报错（如 Exception）信息。(30 分)

（3）数据库中有系统运行后写入的用户信息。(30 分)

3.4　系统日志

系统日志是应用软件中不可缺少的部分，通过日志可以分析判断项目运行过程中在什么位置出现了什么问题。

Apache 的开源项目 Log4j 是一个功能强大的日志组件，通过使用 Log4j 可以控制日志信息输出到控制台 Console、文件、GUI 组件，甚至是套接口服务器、NT 的事件记录器等。也可以控制每一条日志的输出格式，通过定义每一条日志信息的级别，能够更加细致地控制日志的通过一个生成过程。使用配置文件就可以灵活地进行配置，而不需要修改应用程序的代码。

使用 Log4j 实现系统日志的方法如下所述。

（1）添加 log4j-xxx.jar 到 lib 下，已经在 3.2.2 节完成。

（2）在 conf 目录中新建日志配置文件 log4j.properties，内容可以参照互联网上介绍的详细配置，也可以采用下面的简单配置：

```
log4j.rootLogger=debug,Console
log4j.appender.Console=org.apache.log4j.ConsoleAppender
log4j.appender.Console.layout=org.apache.log4j.PatternLayout
log4j.appender.Console.layout.ConversionPattern=%d %l %-5p - %m%n
```

1）log4j.rootLogger 配置根 Logger。包括选择日志输出级别（有 5 个级别可供选择：FATAL、ERROR、WARN、INFO、DEBUG），设置输出位置命名。

2）log4j.appender.appenderName 配置日志信息输出位置 Appender。Appender 为日志输出位置，Log4j 提供的 Appender 有以下几种：

- org.apache.log4j.ConsoleAppender（控制台）。
- org.apache.log4j.FileAppender（文件）。
- org.apache.log4j.DailyRollingFileAppender（每天产生一个日志文件）。
- org.apache.log4j.RollingFileAppender（文件大小到达指定尺寸时产生一个新的文件）。
- org.apache.log4j.WriterAppender（将日志信息以流格式发送到任意指定的地方）。

3）Layout 配置日志输出格式。Log4j 提供的 Layout 有以下几种：

- org.apache.log4j.HTMLLayout（以 HTML 表格形式布局）。
- org.apache.log4j.PatternLayout（可以灵活地指定布局模式）。
- org.apache.log4j.SimpleLayout（包含日志信息的级别和信息字符串）。
- org.apache.log4j.TTCCLayout（包含日志产生的时间、线程、类别等信息）。

4）ConversionPattern 配置打印参数。Log4j 采用类似 C 语言中的 printf 函数的打印格式来格式化日志信息。Log4j 日志输出打印格式见表 3-7。

表 3-7 Log4j 日志输出打印格式

格式	说 明
%m	输出代码中指定的消息
%p	输出优先级，即 DEBUG，INFO，WARN，ERROR，FATAL
%r	输出自应用启动到输出该 log 信息耗费的毫秒数
%c	输出所属的类目，通常就是所在类的全名
%t	输出产生该日志事件的线程名
%n	输出一个回车换行符，Windows 平台为“/r/n”，Unix 平台为“/n”
%d	输出日志时间点的日期或时间，默认格式为 ISO8601，也可以在其后指定格式，比如：%d{yyy MMM dd HH:mm:ss , SSS}，输出格式例：2002 年 10 月 18 日 22:10:28，921
%l	输出日志事件的发生位置，包括类目名、发生的线程，以及在代码中的行数

作为最核心的模块之一，Nutz 的日志信息已经相当完善，看懂 Nutz 的日志，也是了解 Nutz 工作方式的核心途径之一。从控制台 Console 可以看到很多与 Nutz 相关的日志信息，基本包括以下内容：

- NutFilter 启动。
- 资源扫描器开始初始化。
- NutFilter 输出容器信息。
- 加载 MainModule。

- 加载 Ioc 配置信息。
- 解析路径映射的信息。
- 执行用户自定义 Setup 的信息。
- NutFilter 完成的耗时信息。

3.5　用户登录

大部分 Web 系统都具有用户登录模块，因为只有登录用户才能拥有某些特定操作的权限。要实现登录功能，必定有登录页面，当前项目的系统首页与登录页面集成在一个页面上，所以修改了默认的 index.html。

3.5.1　美化系统首页

为了美化系统首页 index.html，如果前端开发能力比较强，建议原创开发一个有特色的首页。为了节省时间，可以从其他网站获取免费开源的页面，例如从模板之家下载，然后可以进行一些修改和内容调整，例如：

- 分离出需要的样式内容，新建 login.css，存放于 WebRoot\include\css。
- 分离出需要的 javascript 内容，新建 login.js，存放于 WebRoot\include\js。
- 整合其他需要的 JS 和 CSS 等文件，放进项目中 include 的相应文件夹。

在 index.html 文件中，引入需要的 JS 和 CSS 等文件，当前页面使用了 EasyUI 框架，所以也引用了 EasyUI 的 JS 和 CSS 等文件。

```
<meta http-equiv="Content-Type" content="text/html; charset=UTF-8" />
<link rel="stylesheet" type="text/css" href="./include/css/main.css" />
<link rel="stylesheet" type="text/css" href="./include/easyui/themes/default/easyui.css" />
<link rel="stylesheet" type="text/css" href="./include/easyui/themes/icon.css" />
<script type="text/javascript" src="./include/js/jquery.min.js"></script>
<script type="text/javascript" src="./include/easyui/jquery.easyui.min.js"></script>
<script type="text/javascript" src="./include/easyui/locale/easyui-lang-zh_CN.js"></script>
<script type="text/javascript" src="./include/js/login.js"></script>
```

- 删除不需要的代码，比如 action，method，type="submit"等。
- 修改部分代码，例如修改用户名和密码 input 的 name 和 id 值，使其与 User 类中的属性一致，以便封装数据传输到后端。

首页核心 html 代码主要实现用户在 Web 系统的前端页面输入用户名和密码，然后单击"登录"按钮。软件开发技术的学习，主要是学会方法，要活学活用，不同的人实现的前端页面可以有不同的风格和特征。首页核心 html 代码如下：

```
<form id="login_form" onsubmit="return false;">
<input type="text" id="uid" placeholder="用户名...">
<input type="password" id="pwd" placeholder="密码..." >
<input id="login_submit" value="登录" >
</form>
```

在 login.js 中添加代码，实现表单信息的获取、检验，并通过 ajax 向后端传输，处理后端返回的消息。以下 JS 代码让首页上的热气球随机移动，呈现动态效果。

```
$(function(){
    airBalloon('div.air-balloon');
});
function airBalloon(balloon){
    var viewSize = [] , viewWidth = 0 , viewHeight = 0 ;
    resize();
    $(balloon).each(function(){
        $(this).css({top: rand(40, viewHeight * 0.5 ) , left : rand( 10 , viewWidth - $(this).width() ) });
        fly(this);
    });
    $(window).resize(function(){
        resize()
        $(balloon).each(function(){
            $(this).stop().animate({top: rand(40, viewHeight * 0.5 ) , left : rand( 10 , viewWidth - 
$(this).width() ) } ,1000 , function(){
                fly(this);
            });
        });
    });
    function resize(){
        viewSize = getViewSize();
        viewWidth = $(document).width();
        viewHeight = viewSize[1] ;
    }
    function fly(obj){
        var $obj = $(obj);
        var currentTop = parseInt($obj.css('top'));
        var currentLeft = parseInt($obj.css('left') );
        var targetLeft = rand( 10 , viewWidth - $obj.width() );
        var targetTop = rand(40, viewHeight /2 );
        var removing = Math.sqrt( Math.pow( targetLeft - currentLeft , 2 )  + Math.pow( targetTop - 
currentTop , 2 ) );
        var moveTime = removing / 24;
        $obj.animate({ top : targetTop , left : targetLeft} , moveTime * 1000 , function(){
            setTimeout(function(){
                fly(obj);
            }, rand(1000, 3000) );
        });
    }
    function rand(mi,ma){
        var range = ma - mi;
        var out = mi + Math.round( Math.random() * range) ;
        return parseInt(out);
    }
```

```
    function getViewSize(){
        var de=document.documentElement;
        var db=document.body;
        var viewW=de.clientWidth==0 ? db.clientWidth : de.clientWidth;
        var viewH=de.clientHeight==0 ? db.clientHeight : de.clientHeight;
        return Array(viewW,viewH);
    }
};
```

3.5.2　Ajax 方法

Ajax 是异步 JavaScript 和 XML，是一种用于创建快速动态网页的技术。Ajax 通过前端页面与后端服务器进行少量数据交换，使网页实现异步更新，可以在不重新加载整个网页的情况下，对网页的某部分进行更新。Ajax 可使 Internet 应用程序更小、更快、更友好。

$.ajax([settings]) 方法通过 HTTP 请求加载远程数据，该方法是 jQuery 底层 Ajax 实现。常用参数 settings 配置如下：

- **url** 发送请求的地址。
- **type** 请求方式 ("POST" 或 "GET")。
- **data** 发送到服务器的数据，必须为 Key/Value 格式。
- **error** 请求失败后的回调函数。
- **success** 请求成功后的回调函数。

采用 jQuery+Ajax，可以非常灵活地校验、封装和提交表单信息。在 login.js 文件中，添加一个 checkUserName 函数，用于校验和提交登录信息，代码如下：

```
Var base="./";
$(function() {
    $('#login_form input').keydown(function (e) {
        if (e.keyCode == 13)
        {
            checkUserName();
        }
});
    $("#login_submit").click(checkUserName);
});

function checkUserName()   //登录前，校验用户信息
{
    var a=$('#uid').val();
    var b=$('#pwd').val();
    if(a==""){alert("请输入用户名");return;}
    if(b==""){alert("请输入登录密码");return;}
    $.ajax({
            url : base+"user/doLogin",
            data:{"uid":a,"pwd":b},   //只封装和传输指定的数据
```

```
                type:"POST",
                success : function (res) {
                    if (res.ok) {
                            window.location.href=base+res.msg;
                    }else {alert(res.msg);}
                    return false;
                },
                error : function(res) {alert("系统错误！");}
        });
    }
```

在运行页面登录窗口的输入框中，按下回车键或直接单击“登录”按钮，则执行函数 checkUserName。在 checkUserName 函数中，首先获取用户输入的用户名和密码，然后校验。如果二者有一个没有输入，则弹出警告框，不再向后端发送登录请求；否则，采用 JSON 形式封装用户名和密码，向后端发送 post 登录请求。后端将处理结果返回，函数中继续判断返回的信息，如果有问题，则弹出提示框，否则跳转到后台管理页面。

3.5.3 更友好的 alert

JS 中的 alert 提示框用于显示一些给用户的提示信息，如图 3-20 所示，使用了语句：alert("请输入用户名");。

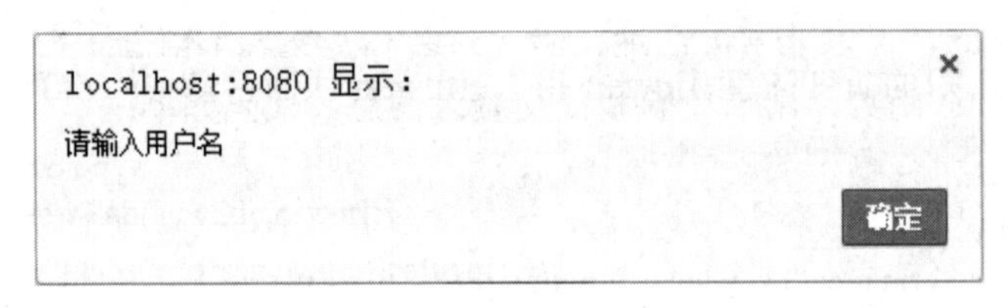

图 3-20 alert 提示框

EasyUI 中的 alert 提示框显示一个提示窗口，比 Js 中的 alert 提示框更加美观，并且可以定制输出，形式更灵活，如图 3-21 所示。

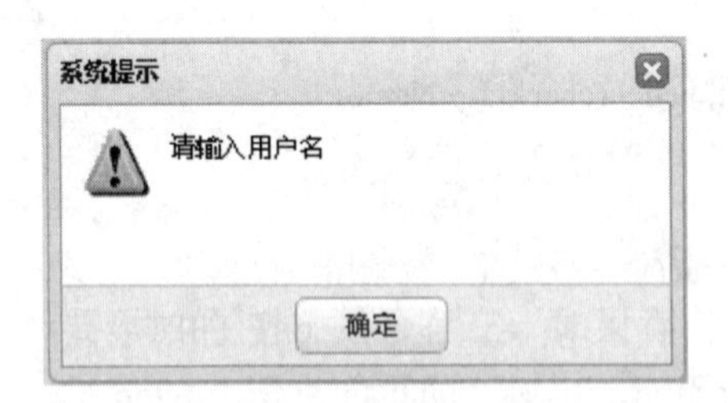

图 3-21 EasyUI 中的 alert 提示窗口

EasyUI alert 的应用方法：

（1）引用 EasyUI 库。在首页中链接 EasyUI 的样式文件和 JS 库文件，通常把第三方官方的 JS 和 CSS 库集中存放在 include 目录下面的专用文件夹，如把 EasyUI 库放在 include 目录下。不同 JS 或者不同 CSS 的引用顺序有先后关系，后引用的样式会代替前面引用的相同样式，但先引用的 JS 变量或函数可以用在后引用的 JS 中。例如 jquery.min.js 的引用比较靠前，因为有很多 JS 会用到 jQuery 技术。引用顺序参考如下：

```
<link rel="stylesheet" href="./include/easyui/themes/default/easyui.css">
<link rel="stylesheet" href="./include/easyui/themes/icon.css">
<script src="./include/easyui/jquery.easyui.min.js"></script>
<script src="./include/easyui/locale/easyui-lang-zh_CN.js"></script>
```

（2）使用语法如下：

```
$.messager.alert(title, msg, icon, fn)
```

参数说明：

title：显示在头部面板上的标题文本。

msg：要显示的消息文本。

icon：要显示的图标图片。可用的值是：**info**、**question**、**error**、**warning**。

fn：当窗口关闭时触发的回调函数。

（3）使用语句如下：

```
$.messager.alert("系统提示","系统提示","请输入用户名","warning");
```

在后面的开发中，均可以使用 EasyUI 的 alert 提示窗口功能。

3.5.4　标题图标

打开某一个网页会在浏览器的标签栏处显示该网页的标题和图标，当网页被添加到收藏夹或者书签中时也会出现网页的图标。添加标题图标会让 Web 系统显得更专业。给网页标题前添加一个小图标，效果如图 3-22 所示。

图 3-22　网页标题与图标

实现方法是用 link 标签实现 shortcut icon，图标文件可以在线制作 ICO，也可用 PhotoShop 制作 256×256 的 png 或 gif 文件，前景和背景对比度明显，前景内容较少但所占空间较多，这样的图标显示效果会比较清晰。制作的图标文件可以存放到 include/img 目录下，通过标签 link 链接。

```
<link rel="shortcut icon" href="./include/img/logo.png">
```

3.5.5　MVC 概述

前端页面设计完成，还需要后端服务器来响应和处理前端页面发送的请求。Nutz.Mvc 需要与 Web 服务器（比如 Tomcat）一起工作，它存在的意义就是要把一个标准的 HTTP 请求转发到某一个 Java 函数中，Nutz.Mvc 的工作方式如图 3-23 所示。用户在浏览器打开 Web 系统前端页面（视图 V），通过 Form 或 Ajax 发送 HTTP 请求，Web 系统通过 MVC 协调，由控制器 C 中特定的方法（Java 函数）来响应用户请求，控制器 C 根据请求，调用模型中的业务逻辑方法，业务逻辑调用 DAO 中的方法访问数据库等持久化数据。模型处理访问的数据后，控制器 C 将处理结果渲染 render，返回 Response 到前端页面，在前端页面显示或自动发起另一个 HTTP 请求。

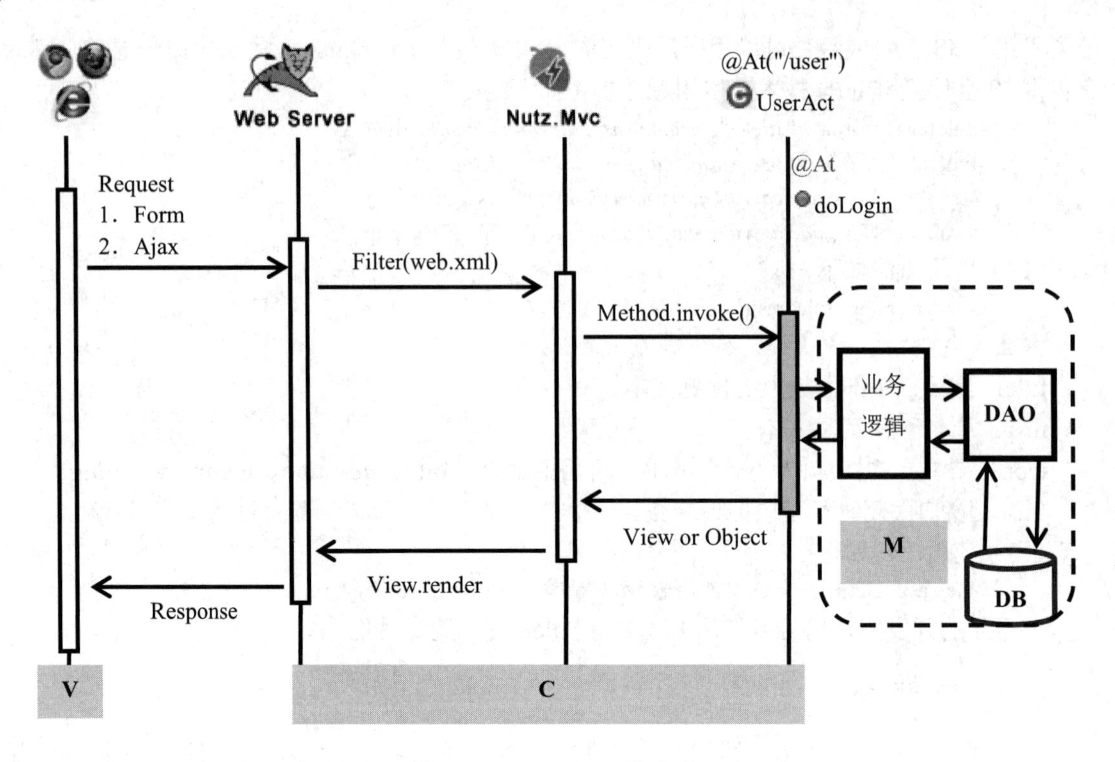

图 3-23　Nutz.Mvc 工作方式

MVC 是指模型（Model）、视图（View）和控制器（Controller），具体描述如下：

（1）Model（模型）表示应用程序核心，是应用程序中用于处理数据逻辑的部分，负责在数据库中存取数据。为了将数据访问与业务逻辑分离，提高业务精度，模型层又划分为 DAO 层和业务层（Service）。

- DAO（Data Access Object）——主要职能是将访问数据库的代码封装起来，让这些代码不会在其他层出现或者暴露出来给其他层。
- 业务层——是整个系统最核心也是最具有价值的一层。该层封装应用程序的业务逻辑，处理用户需求的数据。在业务处理过程中会访问原始数据或产生新数据，DAO 层提供的 DAO 类能很好地帮助业务层完成数据处理。

（2）View（视图）显示数据，通常依据模型数据创建。

（3）Controller（控制器）是应用程序中处理用户交互的部分。通常负责从视图 View 读取用户请求，将其转发给模型 Model，经过处理后把结果返回到视图 View。

MVC 要实现的目标是将软件用户界面 View 和业务逻辑分离，降低代码之间的耦合，以使代码可扩展性、可复用性、可维护性、灵活性加强。使同一个程序可以使用不同的表现形式。控制器 Controller 将用户请求转发给模型层 Model，经过处理后把结果返回到视图 View 层。

3.5.6　MVC 注解

Nutz.Mvc 根据@At 注解将一个 HTTP 请求路径同一个 Java 函数关联起来。子模块中任何函数，只要是 public 且不是 static 的，都可以作为入口函数。标记入口函数的方法是在其上标注注解@At。使用不带参数的@At()注解，默认会使用小写的方法名/类名作为入口 URL；带参数时，以参数作为入口 URL。

任何一个请求，都会经过以下四道工序才能获得返回结果。

（1）过滤：通过 @Filters 注解可以为你的入口函数定义任意多的过滤器

（2）适配：这个过程将 HTTP 输入流转换成入口函数的参数数组，默认输入流是传统的名值对方式，例如文件上传 UploadAdptor。

（3）调用：调用入口函数，函数里面需要调用相关的业务层代码。

（4）渲染：根据入口函数的返回，渲染 HTTP Response。

除了这种标准流程，Nutz.Mvc 还支持“动作链”的方式，可以让用户几乎无限地定制一个请求的处理流程和方式。

如果直接返回一个 View，则用这个 View 来渲染，否则用函数的@Ok 注解声明的 View 来渲染入口函数的返回对象。如果你的函数处理过程中抛出了异常，用@Fail 注解声明的 View 来渲染异常对象，返回值会保存在 request 的 attr 中，名字是 obj。

表 3-8 是子模块上支持的注解，表 3-9 是子模块中入口函数上支持的注解。

表 3-8　子模块上支持的注解

注　解	说　明
@IocBean	AnnotationIocLoader（注解加载器）根据它来判断哪些类应该被自己加载 AnnotationIocLoader 根据@Inject 来注入类的属性（例如 DAO 对象），有@IocBean 它才会生效
@At	模块所有入口函数的 URL 前缀
@Ok	模块默认成功视图
@Fail	模块默认失败视图
@AdaptBy	模块默认 HTTP 参数适配方式
@Filters	模块默认的过滤器数组
@Encoding	模块默认 HTTP 请求的输入输出字符编码

表 3-9　入口函数上支持的注解

注　解	说　明
@At	函数对应的 URL
@Ok	成功视图
@Fail	失败视图
@AdaptBy	HTTP 参数适配方式
@Filters	函数的过滤器数组
@Encoding	HTTP 请求的输入输出字符编码
@GET	限定函数接受 HTTP GET 请求
@POST	限定函数接受 HTTP POST 请求
@PUT	限定函数接受 HTTP PUT 请求
@DELETE	限定函数接受 HTTP DELETE 请求

3.5.7 DAO 接口方法

目前流行的 Web 系统基本都需要数据库的支持，所以在系统的使用过程中，需要访问数据库。软件开发人员需要编写公共 DAO 类，实现添加、删除、修改和查询数据库操作，俗称增删改查（CRUD，即 Create 添加数据、Read 读取查询数据、Update 修改数据和 Delete 删除数据）。传统关系型数据库定义了四种数据操作：插入 Insert、删除 Delete、更新 Update 和查询 Query，这四种操作涵盖了所有的数据操作。Nutz.Dao 已经封装了表 3-10 所示的数据操作。

表 3-10 DAO 的数据操作

操作	方法名	说 明
插入	Insert	一条 SQL 插入一条记录或者多条记录
插入	FastInsert	一条 SQL，通过 batch 插入多条记录
删除	Delete	一条 SQL 删除一条记录
更新	Update	一条 SQL 更新一条或者多条记录
获取	Fetch	一条 SQL 获取一条记录
查询	Query	一条 SQL 根据条件获取多条记录
清除	Clear	一条 SQL 根据条件删除多条记录
建表	Create	根据实体建表
删表	Drop	根据实体/表名称进行删表
聚合	Func	执行 sum、count 等操作

3.5.8 登录方法

在已经了解 Mvc 注解和 DAO 接口方法后，下面介绍实现控制器中的登录方法。在 cn.lrw.newsnutz.module 包中创建子模块 UserAct 类，在 UserAct 类中实现对用户信息的增删改查。例如，用户登录 doLogin，实际上就是查找用户信息。在类中添加 IoC 相关注解或属性，如@IocBean，@Inject，@At 和 DAO 属性。注解@Param 用于获取前端 request 的参数值。doLogin 方法接收登录页面传来的 uid 和 pwd 信息，将其作为查询条件，在 user 表中查找是否有满足条件的用户。根据查询结果，返回 JSON 格式的信息，返回信息使用 NutMap 封装。代码如下：

```
@IocBean
@At("/user")
public class UserAct {
    @Inject
    protected Dao dao;
    //查--登录
    @At
    @Ok("json")
    public Object doLogin(@Param("uid")String uid, @Param("pwd") String pwd, HttpSession
session,HttpServletRequest req){
```

```
            NutMap re = new NutMap();
            User user = dao.fetch(User.class, uid);
            if (user == null ) {
                re.put("ok", false);
                re.put("msg", "用户名不存在");
                return re;
            }
            if (!pwd.equals(user.getPwd())) {
                re.put("ok", false);
                re.put("msg", "密码不正确");
                return re;
            }
            //登录成功后...将用户信息保存到 session
            session.setAttribute("me", user);
            re.put("ok", true);
            re.put("msg", "user/goIndex");
            return re;   //对象
        }

        @At
        @Ok("beetl:web/admin.html")   //用户登录成功，跳转到后台管理页面
        public void goIndex(HttpSession session,HttpServletRequest req){
                User loginuser=(User)session.getAttribute("me");
                req.setAttribute("user", loginuser);
        }
    }
```

3.5.9　匹配视图

通过注解@Ok 和@Fail 可以为入口函数声明不同的视图渲染方式。执行一个业务逻辑可能有两种结果：

- 成功，正常返回。
- 失败，特指入口方法抛出异常。

@Ok 和@Fail 两个注解的值只能是一个字符串，它们的格式是：

"视图类型:视图值"

视图的任务就是将入口函数的返回值（一个 Java 对象）渲染到 HTTP 响应流中。Nutz.Mvc 自带的主要视图类型有：

- JSP：采用 JSP 模板输出网页。
- Redirect：按照给定的视图值发送 HTTP 重定向命令到客户端。
- Forward：服务器端中转，按照给定的视图值，服务器内部重定向到指定的地址。
- JSON：将对象输出成 JSON 字符串。
- void：空白视图，对 HTTP 输出流不做任何处理，空实现。
- Raw：二进制输出，图片输出，文件下载等。

Nutz.Mvc 支持自定义视图规则，例如"beetl:web/admin.html"，指定用模板引擎 Beetl 渲染 HTTP 输出流输出 WEB-INF/web 目录下的 admin.html 页面。

为了验证成功登录，在 WebRoot\WEB-INF\web 目录下，新建一个 admin.html 页面，作为输出的目标视图。

3.5.10 Beetl 配置

在 conf 目录下创建 beetl.properties，添加一条配置项设置根路径，其他的采用默认值。

```
RESOURCE.root=/WEB-INF/
```

Web 项目系统中，除系统首页以外，其他前端页面都可以使用 Beetl 模板引擎渲染，Beetl 远超过主流 Java 模板引擎性能（引擎性能 5～6 倍于 freemaker，2 倍于 JSP），而且消耗较低的 CPU。使用 Beetl 的目的是将模板文件和数据通过模板引擎快速生成最终的 HTML 代码，通过浏览器呈现出来。

3.5.11 退出系统

为了验证退出系统功能，修改 admin.html 页面，在页面上添加一个 EasyUI 按钮，按钮的单击事件为请求退出系统。

1. 引用 JS 和 CSS

```
<link rel="stylesheet" href="${ctxPath}/include/easyui/themes/default/easyui.css">
<link rel="stylesheet" href="${ctxPath}/include/easyui/themes/icon.css">
<script src="${ctxPath}/include/js/jquery.min.js"></script>
<script src="${ctxPath}/include/easyui/jquery.easyui.min.js"></script>
<script src="${ctxPath}/include/easyui/locale/easyui-lang-zh_CN.js"></script>
```

2. 添加按钮

```
<a id="logout" href="#" class="easyui-linkbutton" data-options="iconCls:'icon-cancel'">我要退出系统
</a>
```

3. 添加按钮事件

```
$(function(){
$("#logout").click(function(){
      top.location.href="${ctxPath}/user/doLogout";
});
});
```

4. 全局变量 ctxPath

ctxPath 是 Beetl 中的默认变量，表示 Servlet Context Path，使用方法是用默认占位符号${ }将变量括起来${ctxPath}，代表当前项目在浏览器地址栏中显示的路径，如 http://localhost:8080/newsnutz。

在前端页面中请求/user/doLogout 时，UserAct 子模块中添加响应该请求的方法，退出系统前，必须使 session 中保存的信息失效，然后@Ok(">>:/")跳转到系统首页，代码如下：

```
@At
@Ok(">>:/")
public void doLogout(HttpSession session) {
      session.invalidate();
}
```

3.5.12　密码加密

为了提高开发效率，往往需要使用继承实现代码复用，就是把一个功能写成一个方法，以便当再次需要相同功能的时候可以直接使用，而不用重新编码。在 cn.lrw.newsnutz.utils 包中创建 BaseAct 类，在这个类中封装各种公用的方法。

（1）字符串加密方法。调用 Shiro 中的 SHA 加密算法，可用于密码的不可逆加密。

```
protected String lrwCode(String text,String salt){
    if(salt.equals("")) salt="abcdefghijklmnopqrstuvwx";
    return new Sha256Hash(text, salt,1024).toBase64();
}
```

（2）修改 UserAct 类，使其继承新建的 BaseAct 类。

```
public class UserAct extends BaseAct {
```

在 UserAct 的登录方法 doLogin 中调用加密方法。

```
String p=lrwCode(pwd,"");
if (!p.equals(user.getPwd())) {
```

（3）修改 MainSetup 中的初始化密码。

```
String salt="abcdefghijklmnopqrstuvwx";
user.setPwd(new Sha256Hash("123", salt, 1024).toBase64());
```

（4）删除数据库 user 表中记录。

（5）重新运行项目，将加密的数据写入数据库，测试登录跳转功能。

3.5.13　登录 Filter

为了防止用户不登录系统就能够查看或改动系统中的敏感数据，在用户请求某种操作（登录、注册除外）前，系统自动检查用户是否已经登录，如果没有登录则返回登录页面。为了给用户一个比较明确的提示信息，在 WebRoot\error 目录下新建一个 nologin.html 页面，页面提示登录失效，2 秒钟后跳转到登录页面（系统首页）。

nologin.html 页面 Body 中的内容如下，主要用于提示是否有效登录，读者可以自定义提示页面风格和提示信息。

```
<h1>登录失效</h1>
<p>对不起，您没有登录或者登录已超时。</p>
<script type="text/javascript">
    setTimeout(reDo, 2000);
    function reDo(){ top.location.href = "../";}
</script>
```

在 cn.lrw.newsnutz.utils 包中创建 LoginFilter 类实现接口 ActionFilter，检查用户是否登录（检查当前 Session 是否带有 me 这个 Attribute），从而确定是否允许执行请求的某种操作。

```
public class LoginFilter implements ActionFilter {
    @Override
    public View match(ActionContext context) {
        HttpServletRequest request = context.getRequest();
        User user = (User) request.getSession().getAttribute("me");
```

```
            String contentType = request.getContentType();
            if (user == null) {
                if (Strings.sNull(contentType).contains("application/x-www-form-urlencoded"))
                {
                    context.getResponse().setHeader("sessionstatus", "timeout");
                }
                ServerRedirectView view = new ServerRedirectView("/error/nologin.html");
                return view;
            }
            request.setAttribute("me", user);
            return null;
        }
    }
```

接下来，稍微修改一下 UserAct 类，其中用到@Filters 注解。

- @Filters({@By(type=LoginFilter.class)})，请求执行 UserAct 类中的方法，必须先通过 Filter 中的检验。

```
@Filters({@By(type=LoginFilter.class)})
public class UserAct extends BaseAct{…}
```

- @Filters()，用户登录肯定无法通过 Filter 检验，所以在 doLogin 方法前添加一个无参数的@Filters()，表示临时屏蔽 Filter 检验，允许没有登录的用户请求执行 doLogin 方法。

```
@Filters()
public Object doLogin(…)
```

重新运行项目，查看是否能正常登录跳转；查看退出后在浏览器地址栏输入 http://localhost:8080/newsnutz/user/goIndex，然后回车，能否直接访问后台管理页面 admin.html。这个页面应该是用户登录后才能访问的。如果 Filter 使用正确，由于没有权限直接访问 admin.html，会自动跳转到 error/nologin.html 页面。

3.6 考核任务

（1）具有布局合理且友好的首页。（20 分）
（2）设置标题前的图标。（5 分）
（3）能够正常登录跳转。（40 分）
（4）能够正常退出系统。（10 分）
（5）输入错误的登录信息或者不输入，执行“登录”应有相应明确的提示。（15 分）
（6）友好的弹出信息框。（10 分）

3.7 调试方法

在软件开发过程中，掌握了借助开发工具进行软件调试的方法，就可以通过各种调试方法的应用，更快地定位问题出现的位置，从而有针对性地分析问题、解决问题，软件开发效率能够得到更大程度的提高。

软件开发的难点在于不会调试、不会排（除）错（误）。所以学会各种调试方法，尤其是常用的方法并熟练掌握，在软件开发过程中，就能做到事半功倍，否则会把大量的时间消耗在排错工作上，时间长了，就会丧失信心、失去学习的兴趣和动力。

3.7.1　后端调试

后端调试需要借助开发平台（编译工具），如 MyEclipse、Eclipse、Intellij IDEA 等，采用查看日志和断点跟踪等方法。

1. 查看日志

因为添加了 Log4j 日志系统，Console 控制台会显示很多关于系统启动、运行、停止的日志信息。在项目启动后，通过查看 Console 中的日志或报错（Exception）信息，能解决一部分突出的问题，如语法错误、数据库配置错误等。查看日志的方法是首先把 Console 窗口右边的滚动条拖动到最上方；然后从上往下滑动滚动条，找到第 1 个出现 Exception（通常是蓝色字）的位置，然后根据异常出现位置的上下文，分析异常产生的原因，找到解决异常的方法。有时候会出现很多异常信息，但解决了前面的异常，后面的异常也就不存在了。

有经验的开发调试人员对于常见的异常一看就知道导致异常的原因和解决方法；对于初学者来说，需要积累经验，首先是会用网络资源，然后才是问老师（能者为师），因为网络资源一直都在，而老师不一定能及时响应。

（1）异常样例 1——数据库未启动。

1）现象。图 3-24 所示是控制台显示的日志，日志中提示有 CommunicationsException 异常，原因是 Communications link failure。提示信息表明无法连接数据库。

```
Problems  Console  Servers  Search
2018-09-26 11:04:52,966 org.nutz.dao.jdbc.Jdbcs.<clinit>(Jdbcs.java:90) DEBUG - Jdbcs init complete
2018-09-26 11:04:52,966 org.nutz.dao.jdbc.Jdbcs.getExpert(Jdbcs.java:103) INFO  - Get Connection from DataSource for JdbcE
xpert, if you lock at here, check your database server and configure
2018-09-26 11:04:54,261 com.alibaba.druid.pool.DruidDataSource$CreateConnectionThread.run(DruidDataSource.java:2188) ERROR
 - create connection error, url: jdbc:mysql://127.0.0.1:3066/dbnews1?serverTimezone=Hongkong&useSSL=false, errorCode 0, sta
te 08S01
com.mysql.cj.jdbc.exceptions.CommunicationsException: Communications link failure

The last packet sent successfully to the server was 0 milliseconds ago. The driver has not received any packets from the se
rver.
	at com.mysql.cj.jdbc.exceptions.SQLError.createCommunicationsException(SQLError.java:166)
	at com.mysql.cj.jdbc.exceptions.SQLExceptionsMapping.translateException(SQLExceptionsMapping.java:58)
	at com.mysql.cj.jdbc.ConnectionImpl.createNewIO(ConnectionImpl.java:853)
	at com.mysql.cj.jdbc.ConnectionImpl.<init>(ConnectionImpl.java:440)
	at com.mysql.cj.jdbc.ConnectionImpl.getInstance(ConnectionImpl.java:241)
	at com.mysql.cj.jdbc.NonRegisteringDriver.connect(NonRegisteringDriver.java:221)
	at com.alibaba.druid.pool.DruidAbstractDataSource.createPhysicalConnection(DruidAbstractDataSource.java:1464)
	at com.alibaba.druid.pool.DruidAbstractDataSource.createPhysicalConnection(DruidAbstractDataSource.java:1529)
	at com.alibaba.druid.pool.DruidDataSource$CreateConnectionThread.run(DruidDataSource.java:2185)
```

图 3-24　Console 日志——CommunicationsException 异常

2）分析。造成异常的原因可能是配置文件中使用的数据库没有启动运行。

3）解决。启动数据库，再运行项目。

（2）异常样例 2——数据库配置错误。

1）现象。图 3-25 所示是控制台显示的日志，日志中提示有 SQLException 异常，原因是 Access denied for user 'root'@'localhost' (using password: NO)。提示信息表明数据库拒绝访问。

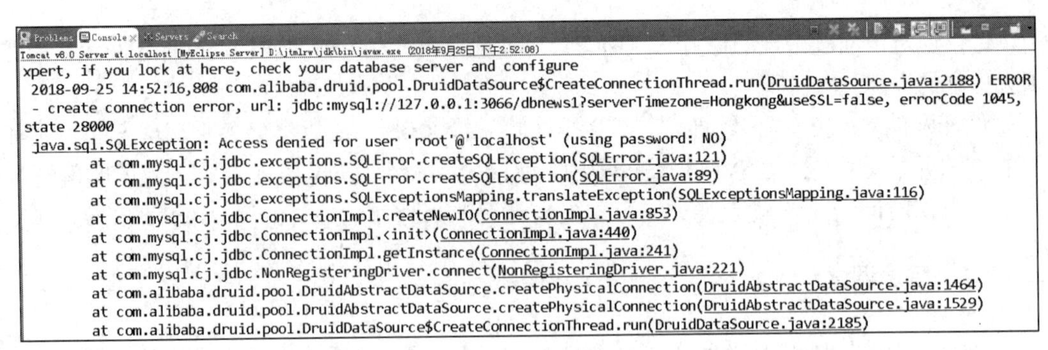

```
Tomcat v8.0 Server at localhost [MyEclipse Server] D:\jtmlrw\jdk\bin\javaw.exe (2018年9月25日 下午2:52:08)
xpert, if you lock at here, check your database server and configure
 2018-09-25 14:52:16,808 com.alibaba.druid.pool.DruidDataSource$CreateConnectionThread.run(DruidDataSource.java:2188) ERROR
 - create connection error, url: jdbc:mysql://127.0.0.1:3066/dbnews1?serverTimezone=Hongkong&useSSL=false, errorCode 1045,
state 28000
 java.sql.SQLException: Access denied for user 'root'@'localhost' (using password: NO)
	at com.mysql.cj.jdbc.exceptions.SQLError.createSQLException(SQLError.java:121)
	at com.mysql.cj.jdbc.exceptions.SQLError.createSQLException(SQLError.java:89)
	at com.mysql.cj.jdbc.exceptions.SQLExceptionsMapping.translateException(SQLExceptionsMapping.java:116)
	at com.mysql.cj.jdbc.ConnectionImpl.createNewIO(ConnectionImpl.java:853)
	at com.mysql.cj.jdbc.ConnectionImpl.<init>(ConnectionImpl.java:440)
	at com.mysql.cj.jdbc.ConnectionImpl.getInstance(ConnectionImpl.java:241)
	at com.mysql.cj.jdbc.NonRegisteringDriver.connect(NonRegisteringDriver.java:221)
	at com.alibaba.druid.pool.DruidAbstractDataSource.createPhysicalConnection(DruidAbstractDataSource.java:1464)
	at com.alibaba.druid.pool.DruidAbstractDataSource.createPhysicalConnection(DruidAbstractDataSource.java:1529)
	at com.alibaba.druid.pool.DruidDataSource$CreateConnectionThread.run(DruidDataSource.java:2185)
```

图 3-25 Console 日志——SQLException 异常

2）分析。造成异常的原因可能有：配置文件中数据库名称错误、访问数据库的用户名或密码有错误。

3）解决。打开 conf/ioc/dao.js，检查并修改其中数据库的配置信息。

（3）异常样例 3——Java 方法错误。

1）现象。图 3-26 所示是控制台显示的日志，日志中提示有 Error，原因是 The method lrwcode(String, String) is undefined for the type UserAct，提示信息表明存在编译错误，在 UserAct 类中未定义方法 lrwcode(String,String)。日志中还显示 Error 产生的位置信息：UserAct.java 文件的第 35 行，处于 doLogin 方法内部。

```
Tomcat v8.0 Server at localhost [MyEclipse Server] D:\jtmlrw\jdk\bin\javaw.exe (2018年9月26日 下午12:23:20)
 2018-09-26 12:24:22,852 org.nutz.ioc.impl.NutIoc.get(NutIoc.java:165) DEBUG - Get 'dao'<interface org.nutz.dao.Dao>
 2018-09-26 12:24:22,854 org.nutz.mvc.impl.processor.FailProcessor.process(FailProcessor.java:28) WARN  - Error@/user/doLog
in :
 java.lang.Error: Unresolved compilation problem:
	The method lrwcode(String, String) is undefined for the type UserAct

	at cn.lrw.newsnutz.module.UserAct.doLogin(UserAct.java:35)
	at cn.lrw.newsnutz.module.UserAct$FM$doLogin$1e743dd9252e6cb7161c1420ba1b26b3.invoke(UserAct.java)
	at org.nutz.mvc.impl.processor.MethodInvokeProcessor.process(MethodInvokeProcessor.java:31)
	at org.nutz.mvc.impl.processor.AbstractProcessor.doNext(AbstractProcessor.java:44)
	at org.nutz.mvc.impl.processor.AdaptorProcessor.process(AdaptorProcessor.java:30)
	at org.nutz.mvc.impl.processor.AbstractProcessor.doNext(AbstractProcessor.java:44)
	at org.nutz.mvc.impl.processor.ActionFiltersProcessor.process(ActionFiltersProcessor.java:58)
	at org.nutz.mvc.impl.processor.AbstractProcessor.doNext(AbstractProcessor.java:44)
	at org.nutz.mvc.impl.processor.ModuleProcessor.process(ModuleProcessor.java:123)
	at org.nutz.mvc.impl.processor.AbstractProcessor.doNext(AbstractProcessor.java:44)
	at org.nutz.mvc.impl.processor.EncodingProcessor.process(EncodingProcessor.java:27)
	at org.nutz.mvc.impl.processor.AbstractProcessor.doNext(AbstractProcessor.java:44)
	at org.nutz.mvc.impl.processor.UpdateRequestAttributesProcessor.process(UpdateRequestAttributesProcessor.java:15)
	at org.nutz.mvc.impl.NutActionChain.doChain(NutActionChain.java:44)
```

图 3-26 Console 日志——lang.Error

2）分析。造成这种异常的原因可能有：方法名有误，方法参数有误或者确实没有定义过提示信息中的方法。

3）解决。如果方法有误，请检查并修改；如果没有定义，请在相应类中添加方法的定义。

不同的开发人员、同一开发人员在不同开发场景，一般都会遇到异常或错误，无法列举所有可能的问题，所以需要开发人员在开发过程中，不断积累经验。

2. 断点跟踪

在 Java 编写的方法中，可以在代码行前面添加断点。以 debug 模式运行项目，运行到断点时，可以单步跟踪调试，可以查看前端传来的参数值，可以查看业务处理后的属性值，可以查看业务处理过程中的异常信息等。学会断点跟踪可以帮助分析问题可能出现在代码的什么位置及原因，最终根据原因找出解决方案。

注意：不要在 MainSetup 的 init 方法中加断点，这会导致 debug 模式启动超过 45s。

单步调试有三种方法，根据实际情况选择单步调试方法，在 MyEclipse 的工具栏上有相应的 3 个按钮及快捷键。

- Step Into（F5）：单步执行，如果该行有自定义方法，则运行进入自定义方法（不会进入官方类库的方法）。
- Step Over（F6）：单步执行，如果当前行有方法调用，这个方法将被执行完毕返回，然后到下一行。
- Step Return（F7）：当单步执行到一个方法的内部，但觉得该方法没有问题，就可以使用 Step Return 执行完当前方法的余下部分，并返回到该方法被调用处的下一行语句。

如果想直接执行到下一个断点，单击 Resume，程序将运行当前断点到下一个断点之间需要执行的代码。如果后面代码没有断点，再次单击该按钮将会执行完程序。

如果不想再继续，单击 Terminate 停止项目的 debug 运行。

以用户登录为样例，如图 3-27 所示，在 doLogin 方法内的第 1 行的行号（28）左边空白处，双击添加一个断点（BreakPoints、双击断点可以取消当前断点）。断点的添加，可以在项目运行前、也可以在项目运行后。

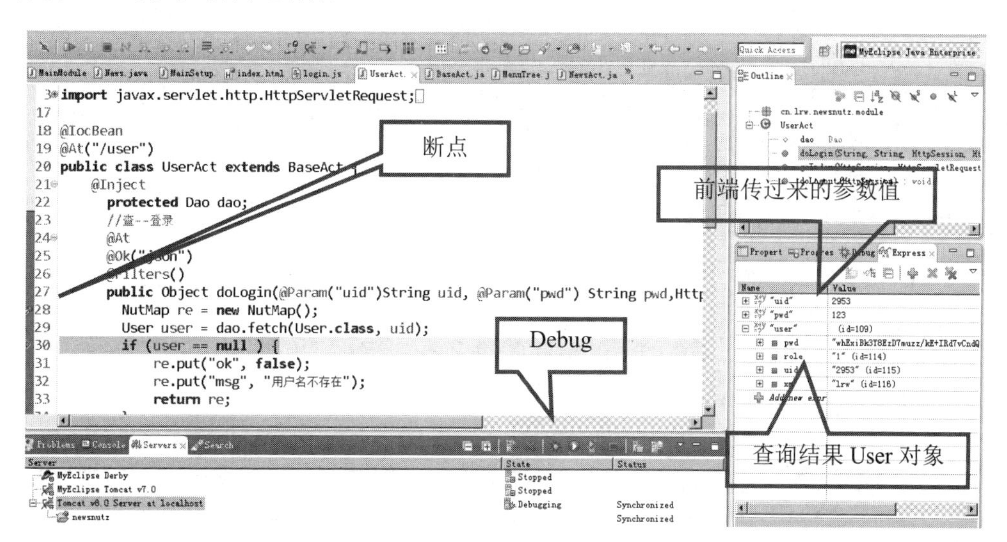

图 3-27　后端断点跟踪

以 debug 模式运行当前项目，在浏览器中打开系统的登录页，输入用户名和密码后登录到系统；当运行到断点时，断点处显示有一个向左的箭头，表示程序执行到当前行已暂停；按两次热键 F6，程序执行到第 30 行。单击菜单 Window 项，选择 Show View 中的 Expressions 命令，打开 Expressions 表达式视图，输入需要查看的属性名、参数名或对象名，就可以看到它们当前的值。

如果查看的值，不是预计的值，则要往回查；如果是没有取到前端传过来的参数值，则甚至要检查前端的业务逻辑。

3.7.2 前端调试

前端调试需要借助浏览器的开发者工具，如 DevTool、Firebug 或者 Firefox DevTools 等。建议使用 360 极速浏览器的极速模式，其他国产浏览器的极速模式、Chrome 浏览器等也可以。

1. 断点跟踪

运行项目，确保控制台没有异常、没有错误。在浏览器中打开项目登录页面，如图 3-28 所示。调出浏览器的“开发者工具”（大部分浏览器通常用快捷键 F12），选择 Sources 面板，展开左边区域树形资源，选择 include/js/login.js，在中间区域 login.js 的 javascript 代码单击行号 98，即可以添加一个断点，如果单击断点，则是取消该断点。展开右边区域 Watch，添加（输入）需要查看的变量、表达式或对象，则可以看到它们在程序执行到断点时的值。可以单步调试、也可以继续执行到下一断点。

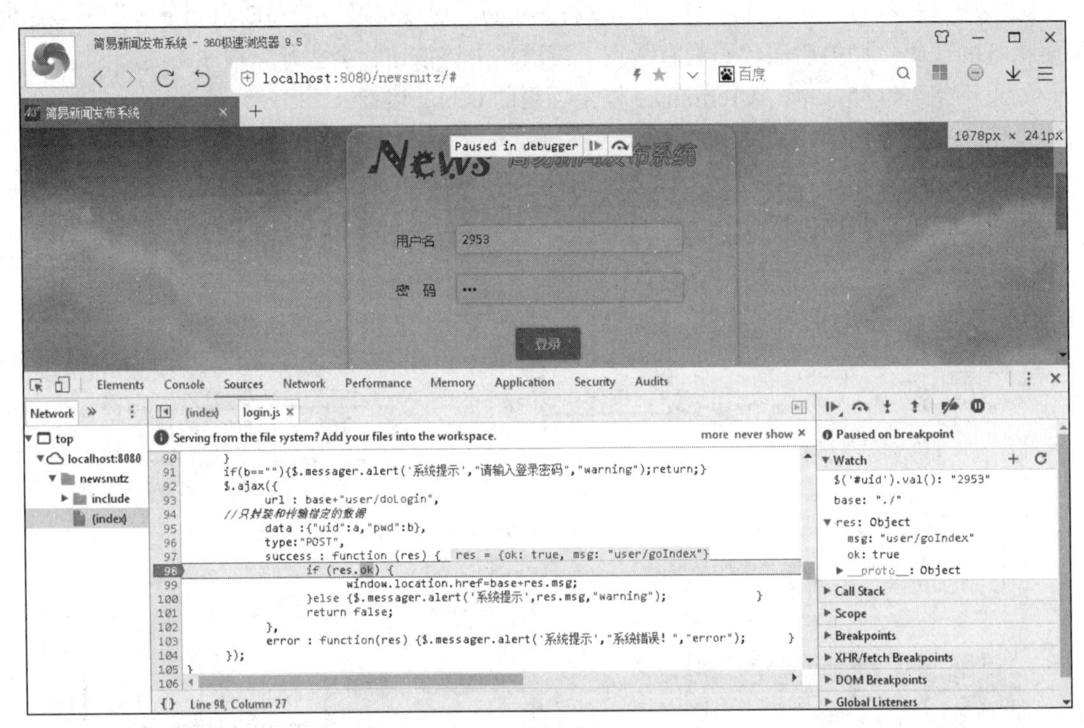

图 3-28 前端断点跟踪

如果查看的值，不是预计的值，则要往回查；如果是没有取到后端传过来的值，可能需要检查后端的业务逻辑。

2. 查看代码与 CSS

在浏览器开发者工具中打开 Elements 面板，可以查看、查找网页源代码 HTML 中的任一元素及其用到的 js/css/image 文件等，如果手动修改任一元素的属性和样式则能实时在浏览器里面得到反馈，如图 3-29 所示，当前选中了 id 为 login_form 的 div 元素，右边显示了这个元素上的 CSS，可以临时手工调整它的 CSS 属性和属性值，查看即时反馈的 UI 界面效果，如果觉得合适，就参照临时修改后的 CSS，修改或添加到源文件中。

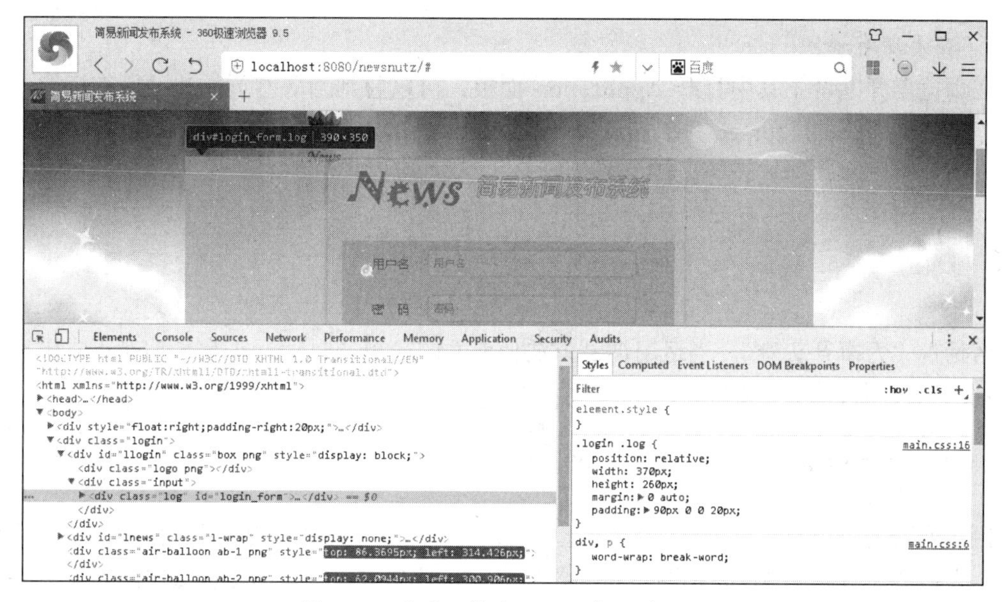

图 3-29　查看、修改 HTML 代码或 CSS

3. 查看网络请求信息

在浏览器开发者工具中打开 Network 面板，从发起网页页面请求 Request 后，分析 HTTP 请求后得到的各个请求资源信息（包括状态、资源类型、大小、所用时间等），可以根据这个进行网络性能优化。选择一个向后端发起的请求，例如登录 doLogin，则可以查看请求时的 Headers 消息和 Response 的数据消息，如图 3-30 所示，这是后端服务器响应 Response 的消息。

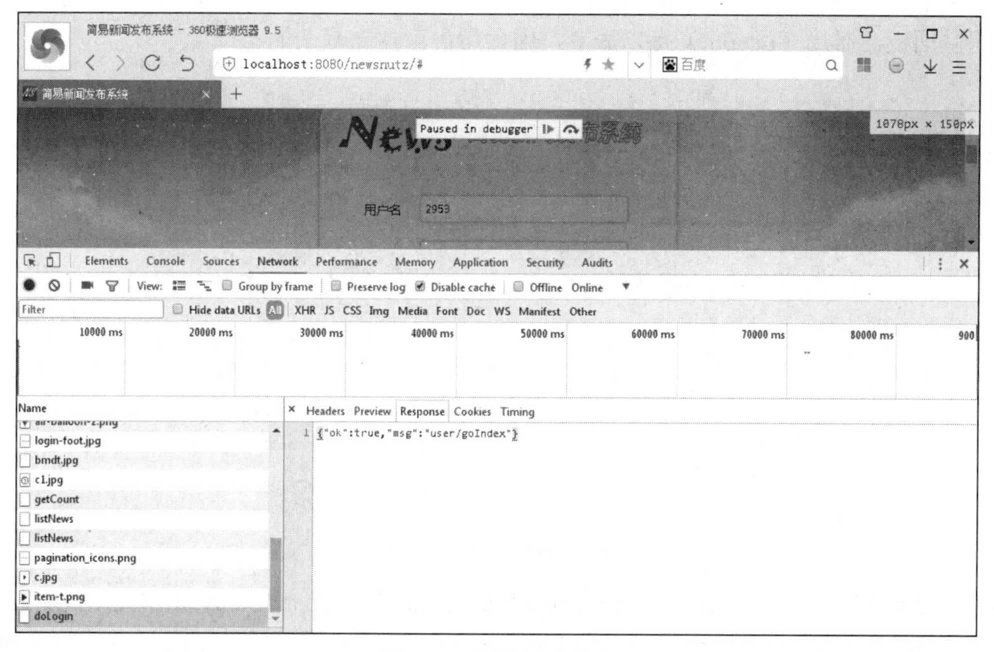

图 3-30　网络请求信息

4. 查看所有资源

在浏览器开发者工具中打开 Application 面板，可以看到当前网页加载的所有资源信息，包括存储数据（Local Storage、Session Storage、IndexedDB、Web SQL、Cookies）、缓存数据、字体、图片、脚本、样式表等。

3.8 新闻管理

俗话说“万事开头难”，当运用调试方法使前面的简单项目运行正常、功能正常以后，会更顺利地完成项目的其他功能。

3.8.1 后台 Layout

网页布局是网站前端设计的重中之重，如何把文字、图片等网页元素有规则地排列在网页中，达到良好的视觉效果，是网页布局要考虑的重要方面。网站选择什么样的布局直接影响到访客在浏览器上看到的整体效果，页面设计得好与不好，直接影响到访客在网站的停留时间长短。

好的网页布局不仅具有好的页面效果，而且能够让前端开发人员更好地把握网页的整体结构，提高代码的书写效率，具有复用性和可维护性。

网页布局有很多种，EasyUI 提供的布局是比较通用的布局模式，通过组合可以演变成很多种流行的布局形式。如图 3-31 所示，EasyUI 布局（Layout）是有五个区域（北区 North、南区 South、东区 East、西区 West 和中区 Center）的容器。中间的区域面板是必需的，边缘区域面板是可选的。每个边缘区域面板可通过拖拽边框调整尺寸，也可以通过单击折叠触发器来折叠面板。布局（Layout）可以嵌套，因此用户可建立复杂的布局。

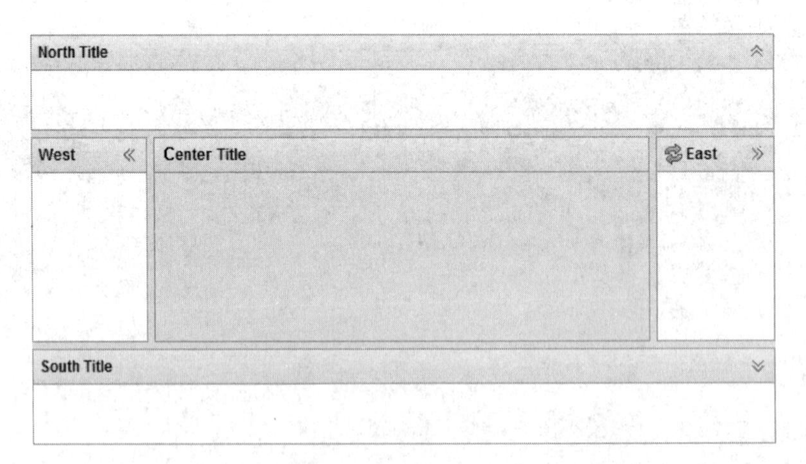

图 3-31 EasyUI 布局（Layout）

修改前面创建的简单后台管理页 admin.html，添加 EasyUI 布局 Layout 和 EasyUI Tabs 选项卡。如图 3-32 所示，在后台管理页面使用默认布局中的 4 个区域（North、West、South 和 Center），分别在不同的区域放置不同类型的内容，通用的做法如下所述。

- North 区域放置系统 LOGO、已登录用户的名称、“退出”按钮或链接。
- West 区域加载树形菜单。
- South 区域显示版权信息。
- Center 区域使用 Tab 来展示内容窗口。

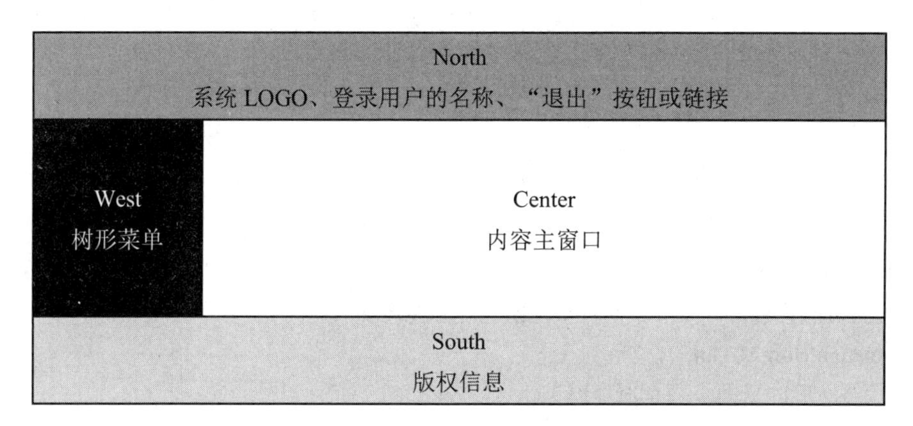

图 3-32　后台布局示意图

实现上述布局的参考代码如下：

```
<body class="easyui-layout">
    <div data-options="region:'north',border:false"
            style="background:#B3DFDA;padding:0 10px 0 10px;vertical-align: middle;">
            <img src="${ctxPath}/include/img/logo.png" width="126" height="50" />
            <div style="float:right;line-height:50px;margin-right:10px;">
                <a id="logout" href="#" class="easyui-linkbutton" data-options=
                    "iconCls:'icon-cancel'">退出</a>
            </div>
            <div style="float:right;line-height:50px;margin-right:10px;">登录用户：${me.xm}
                |</div>
    </div>
    <div data-options="region:'west',split:true,title:'系统菜单'" style="width:180px;padding:10px;">
            <ul id="menutree" class="easyui-tree"></ul>
    </div>
    <div data-options="region:'south',border:false" style="height:30px;padding:5px; text-align:center;
        font-family: arial;color: #A0B1BB;">Copyright © 2017 JavaEE. All Rights Reserved.
    </div>
    <div data-options="region:'center'">
        <div id="tabs" class="easyui-tabs" fit="true" border="false">
        </div>
    </div>
</body>
```

3.8.2　Tab 操作

Tab 选项卡用于在固定大小的区域显示多个选项卡，通过切换选项卡可以看到每个选项卡呈现的不同内容。图 3-33 所示是 EasyUI 提供的 Tab 控件。

为了能够在布局的 Center 区域，根据所选菜单栏目，动态加载相应内容，在 include\js\中创建 lrwtab.js 文件，在文件中自定义三个 JS 函数，实现关闭所有 Tab、切换 Tab、添加 Tab 的功能，然后在 admin.html 页面中引用 lrwtab.js 文件。

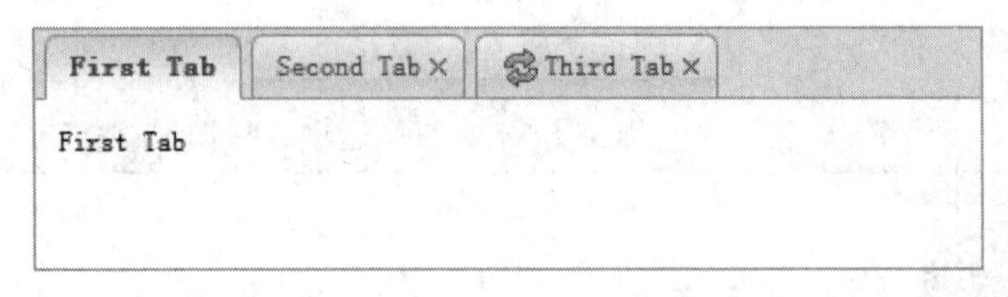

图 3-33 EasyUI Tab 选项卡

1. 关闭所有 Tab

closeAllTabs 函数首先获取所有的 Tab，然后逐个关闭。

```
function closeAllTabs(){
    var tabs=$("#tabs").tabs("tabs");
    for(var i=0;i<tabs.length;i++){
        $("#tabs").tabs("close",i);
    }
}
```

2. 切换 Tab

swNewTab 函数首先判断要切换的 Tab 是否存在，如果存在则保持不变，否则关闭所有 Tab，再添加一个新 Tab，标题名称为 newtitle。将 newurl 请求的页面以 iframe 形式嵌入内容窗口。

```
function swNewTab(newtitle,newurl){
    if($('#tabs').tabs('exists',newtitle))return;
    closeAllTabs();
    $('#tabs').tabs('add',{
        title:newtitle,
        content:'<iframe id="mainframe" name="mainframe" scrolling="auto" height="99%" width="99%" frameboder="0" src="'+newurl+'"></iframe>',
        closable:true
    });
}
```

3. 添加 Tab

addNewTab 函数首先判断要添加的 Tab 是否存在，如果存在则保持不变，否则添加一个新 Tab，标题名称为 newtitle。将 newurl 请求的页面，以 iframe 形式嵌入内容窗口。

```
function addNewTab(newtitle,newurl){
    if($('#tabs').tabs('exists',newtitle))return;
    $('#tabs').tabs('add',{
        title:newtitle,
        content:'<iframe id="mainframe" name="mainframe" scrolling="auto" height="99%" width="99%" frameboder="0" src="'+newurl+'"></iframe>',
```

```
            closable:true
        });
    }
```

3.8.3　封装 Tree 型数据

在 cn.lrw.newsnutz.pojo 中创建 EasyUITree 类，根据 EasyUI 中 Tree 控件的属性，设置 EasyUITree 对象的属性。

```
public class EasyUITree{
    private String id;
    private String text;
    private Boolean checked = false;
    private Map<String,Object> attributes;
    private String state = "closed";
    public List<EasyUITree> children;
    //省略了部分 getter/setter 方法
    public List<EasyUITree> getChildren() {
            return children;
    }
    public void setChildren(List<EasyUITree> children) {
            this.children = children;
    }
}
```

在 cn.lrw.newsnutz.utils 中创建 MenuTree 类，添加 menutree 方法。该方法是从数据库的 cmenu 表中取出菜单栏目数据，然后封装成 JSON 格式的 EasyUITree 数据。如果登录的用户角色不在菜单的 permission 中，则不会封装到返回数据中，用户在前端页面也就看不到该菜单栏目。在教学中菜单栏目已适当简化，层级最多 2 级且栏目的顺序不能动态调整。实现 menutree 方法的代码如下：

```
@At
@Ok("raw")
public String menutree(HttpServletRequest req, HttpSession session) {
    User user = (User) session.getAttribute("me");
    String role=user.getRole();
    //父级菜单
    List<Cmenu> menulist=dao.query(Cmenu.class, Cnd.where("pid","=",0).asc("id"));
    List<EasyUITree> eList = new ArrayList<EasyUITree>();
    if(menulist.size() != 0){
        for (int i = 0; i < menulist.size(); i++) {
            Cmenu t = menulist.get(i);
            if(!t.getPermission().contains(role))continue;
            EasyUITree e = new EasyUITree();
            e.setId(t.getId());
            e.setText(t.getName());
```

```
                List<EasyUITree> eList2 = new ArrayList<EasyUITree>();
                List<Cmenu> menu2 = dao.query(Cmenu.class, Cnd.where("pid", "=", t.getId()).asc("id"));
                for (int j = 0; j < menu2.size(); j++) {//二级菜单
                    Cmenu t2 = menu2.get(j);
                    if(!t2.getPermission().contains(role))continue;
                    Map<String,Object> attributes = new HashMap<String, Object>();
                    attributes.put("url", t2.getUrl());
                    attributes.put("role", t2.getPermission());
                    EasyUITree e1 = new EasyUITree();
                    e1.setAttributes(attributes);
                    e1.setId(t2.getId()+"");
                    e1.setText(t2.getName());
                    e1.setState("open");
                    eList2.add(e1);
                }
                e.setChildren(eList2);
                e.setState("closed");
                eList.add(e);
            }
        }
        return Json.toJson(eList);
    }
```

3.8.4 加载 Tree 型菜单栏目

使用代码$("#menutree").tree()在网页左侧（West）动态加载树形菜单，菜单数据来源于后台以 JSON 形式封装的层级菜单数据。

如果单击一个菜单项，则在 Tab 中打开相应的页面 swNewTab(node.text,“${ctxPath}”+node.attributes.url)，并在标签上面显示相应的名称 node.text。如果单击一个父级菜单，则折叠（Collapse）其他打开的父级菜单，然后展开（Expand）当前单击的父级菜单，显示其下属的二级菜单项。

上述操作的实现代码如下：

```
var opened_node;
$("#menutree").tree(
{
    url : "${ctxPath}/menutree",
    animate : true,
    onClick : function(node) {
        if (!node.attributes) {
            if (!opened_node) {
                $("#menutree").tree('expand', node.target);
                opened_node = node.target;
            } else if (opened_node != node.target) {
```

```
                    $("#menutree").tree('collapse', opened_node);
                    $("#menutree").tree('expand', node.target);
                    opened_node = node.target;
                }
            } else {
                swNewTab(node.text,"${ctxPath}" +node.attributes.url);
            }
        },
        onLoadSuccess : function(node, data) {
            $("#menutree").tree('expandAll');
        }
    });
```

运行项目，登录后台查看，是否在布局左边正常显示菜单栏目，如图 3-34 所示。

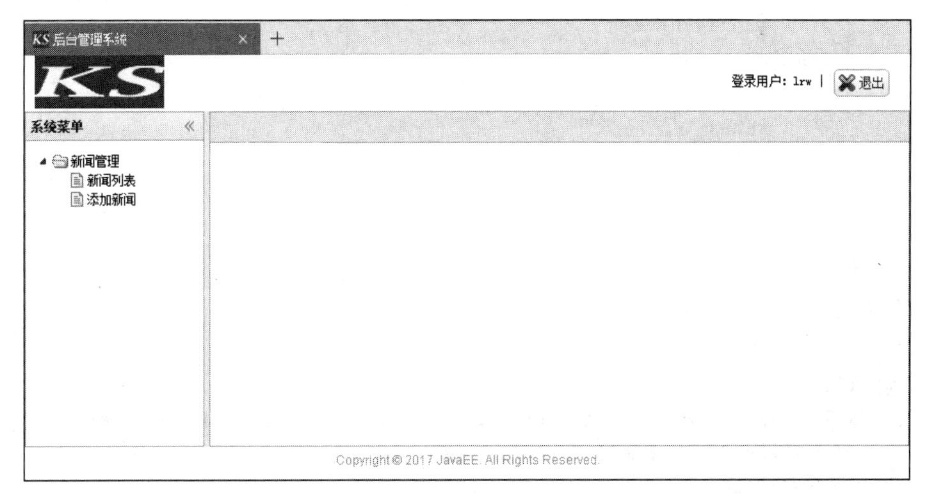

图 3-34 Tree 型菜单

3.8.5 后端新闻业务逻辑

在 cn.lrw.newsnutz.module 包中，创建新闻信息的业务逻辑处理类 NewsAct 可能要用到公有方法，则应继承 BaseAct 类。在类名前添加注解@IocBean、类的入口注解@At("/news")、登录过滤器@Filters({@By(type=LoginFilter.class)})，注入 dao 对象，添加对新闻的增删改查业务处理方法。代码如下：

```
@IocBean
@At("/news")
@Filters({@By(type=LoginFilter.class)})
public class NewsAct extends BaseAct {
    @Inject
    protected Dao dao;
```

注意：①提交新闻的日期时间在服务器生成，避免由于客户端的设备时间有误带来日期时间的不一致性；②只要用户请求阅读一条新闻，则将该新闻的阅读量增加 1。

1．添加新闻

（1）请求跳转到添加新闻的页面 newsadd.html。

```
@At
@Ok("beetl:web/newsadd.html")
public void goAdd(HttpSession session,HttpServletRequest req){
}
```

（2）保存新闻。接收用户在前端页面 newsadd.html 添加的新闻信息，参数@Param("..")中的".."表示对象，补充发布时间和阅读量（默认为 0），然后保存到数据库。

```
@At
@Ok("raw")
public String saveAdd(@Param("..")News news,HttpSession session,HttpServletRequest req){
    try{
            news.setTjdate(new Date());
            news.setHitnum(0);
            dao.insert(news);
    }catch(Exception e){
            e.printStackTrace();
            return "添加失败";
    }
    return "true";
}
```

2．删除新闻

删除指定 id 的一条新闻。

```
@At
@Ok("raw")
public String doDel1(@Param("id")int id,HttpSession session,HttpServletRequest req){
    if(dao.delete(News.class,id)>0)return "true";
    return "false";
}
```

3．修改新闻

（1）请求跳转到修改新闻的页面 newsedit.html，同时从数据库获取指定 id 的新闻信息，通过 request 带回前端页面。

```
@At
@Ok("beetl:web/newsedit.html")
public void goEdit(@Param("id")int id,HttpSession session,HttpServletRequest req){
    News news=dao.fetch(News.class,id);
    req.setAttribute("news", news);
}
```

（2）保存修改后的新闻。接收用户在前端页面 newsedit.html 修改的新闻信息，更新到数据库。

```
@At
@Ok("raw")
public String saveEdit(@Param("..")News news,HttpSession session,HttpServletRequest req){
    try{
```

```
                news.setTjdate(null);//
                if(dao.updateIgnoreNull(news)==1)return "true";
                else return "修改失败";
            }catch(Exception e){
                e.printStackTrace();
                return"修改失败";
            }
        }
```

4. 查询新闻

如果允许没有登录后台的用户查看新闻，需要在相应方法前面添加注解@Filters()，暂时屏蔽类前的 LoginFilter，使得调用该方法不需要有已登录用户的信息。

```
        @At
        @Ok("raw")   //新闻数量
        @Filters()
        public int getCount(HttpSession session,HttpServletRequest req){
            return dao.count(News.class);
        }
        @At
        @Ok("beetl:web/newslist.html")   //跳转到新闻信息列表
        public void goList(HttpSession session, HttpServletRequest req){
        }
        @At
        @Ok("beetl:web/newsread.html")   //阅读指定 id 的新闻内容
        @Filters()
        public void getNews(@Param("id")int id,HttpSession session,HttpServletRequest req){
            News news = dao.fetch(News.class, id);
            int hitnum=news.getHitnum()+1;
            news.setHitnum(hitnum);
            dao.updateIgnoreNull(news);
            req.setAttribute("news", news);
            }
        @At
        @Ok("raw")      //分页查询指定新闻或所有新闻
        @Filters()
        public String listNews(@Param("page") int curPage, @Param("rows") int pageSize, @Param("s_name")
        String s_name, HttpSession session) {
            Criteria cri = Cnd.cri();
            if (!Strings.isBlank(s_name)) {
                cri.where().andLike("title", s_name);
            }
            else cri.where().andEquals("1", 1);
            cri.getOrderBy().desc("id");
            return listPageJson(dao, News.class, curPage,pageSize, cri);
        }
```

3.8.6 封装 DataGrid 数据

为了能让前端 EasyUI DataGrid 获取正确的数据，在 BaseAct 中添加 listPageJson 方法，从数据库按查询条件 cnd 获取数据后，再次封装成包含当前页码 curPage、每页新闻数量 pageSize、新闻列表 List<News>等信息的 JSON 格式数据。

```
public <T> String listPageJson(Dao dao, Class<T> obj, int curPage, int pageSize, Condition cnd) {
        Map<String, Object> jsonobj = new HashMap<String, Object>();
        Pager pager = dao.createPager(curPage, pageSize);
        List<T> list = dao.query(obj, cnd, pager);
        pager.setRecordCount(dao.count(obj, cnd));
        jsonobj.put("total", pager.getRecordCount());
        jsonobj.put("rows", list);
        return Json.toJson(jsonobj);
}
```

3.8.7 后端文件上传

由于前端发布新闻时可能需要上传文件，比如图片、附件等，后端必须有相应的响应方法，将上传的文件写到服务器上指定的文件夹中。

在 cn.lrw.newsnutz.utils 中创建 FileAct 类，封装对上传文件数据的获取、生成新文件名、按上传日期创建文件夹分类存储到服务器磁盘中的操作。

前端拟采用百度编辑器，所以上传文件存储到服务器上的方法，只针对此编辑器，存储的具体位置由配置文件指定。其他在线编辑器，有相似的使用方法，可以参考百度编辑器，灵活使用。

```
@IocBean
@At("/file")
@Filters({@By(type=LoginFilter.class)})
public class FileAct {
        @At
        @Ok("raw")
        public void bdupfile(HttpSession session,HttpServletRequest req,HttpServletResponse res)throws
IOException{
                req.setCharacterEncoding( "UTF-8" );
                res.setHeader("Content-Type" , "text/html");
                String rootPath = Mvcs.getServletContext().getRealPath( "/" );
                PrintWriter out = res.getWriter();
                out.write( new ActionEnter( req, rootPath ).exec() );
        }
}
```

3.8.8 修改 UEditor1.5

从官网下载 UEditor1.5.0 开发版源码。在 MyEclipse 中新建一个 Java Project，假定命名为 ueditor1.5.0，如图 3-35 所示。把 UEditor1.5.0 源码包中 jsp/src 里面的 package 包放进项目

ueditor1.5.0 的 src 中，修改 com.baidu.ueditor 包中的 ConfigManager.java 文件，在它的 initEnv 方法中添加图 3-35 右侧所示框中的代码，这些代码影响配置文件 config.json 中上传文件的路径。导出项目为 jar 文件，假定为 ueditor1.5.0.jar。

```
169 private void initEnv () throws FileNotFoundException, IOException {
170     File file = new File( this.originalPath );
171     if ( !file.isAbsolute() ) {
172         file = new File( file.getAbsolutePath() );
173     }
174     this.parentPath = file.getParent();
175     String configContent = this.readFile( this.getConfigPath() );
176     try{
177         JSONObject jsonConfig = new JSONObject( configContent );
178         ///////下面是手动修改配置文件的前缀////////////////////////////////
179         if(!StringUtils.isEmpty(this.contextPath)){
180             jsonConfig.put("imageUrlPrefix",this.contextPath);//图片
181             jsonConfig.put("scrawlUrlPrefix",this.contextPath);//涂鸦
182             jsonConfig.put("snapscreenUrlPrefix",this.contextPath);//截图
183             jsonConfig.put("catcherUrlPrefix",this.contextPath);//抓取远程图片配置
184             jsonConfig.put("videoUrlPrefix",this.contextPath);//上传视频配置
185             jsonConfig.put("fileUrlPrefix",this.contextPath);//上传文件
186             jsonConfig.put("imageManagerUrlPrefix",this.contextPath);//列出指定目录下的图片
187             jsonConfig.put("fileManagerUrlPrefix",this.contextPath);//列出指定目录下的文件
188         }
189         ////////////////////////////////////////////////////////////////
190         this.jsonConfig = jsonConfig;
191     } catch ( Exception e ) {
192         this.jsonConfig = null;
193     }
194 }
```

图 3-35　修改 ueditor1.5.0

3.8.9　后台新闻信息处理

实现后台添加、删除、修改和查询新闻信息的前端页面。

1. *后台新闻列表*

在 WebRoot\WEB-INF\web 中创建 newslist.html 页面，使用 DataGrid 呈现新闻列表信息，在页面上可以查询新闻、可以请求跳转到修改新闻的页面、可以请求删除新闻，如图 3-36 和图 3-37 所示。引入的 JS 和 CSS 样式文件，基本与 admin.html 相同，所以可以用复制页面、修改页面的方法，提高页面的创建效率。

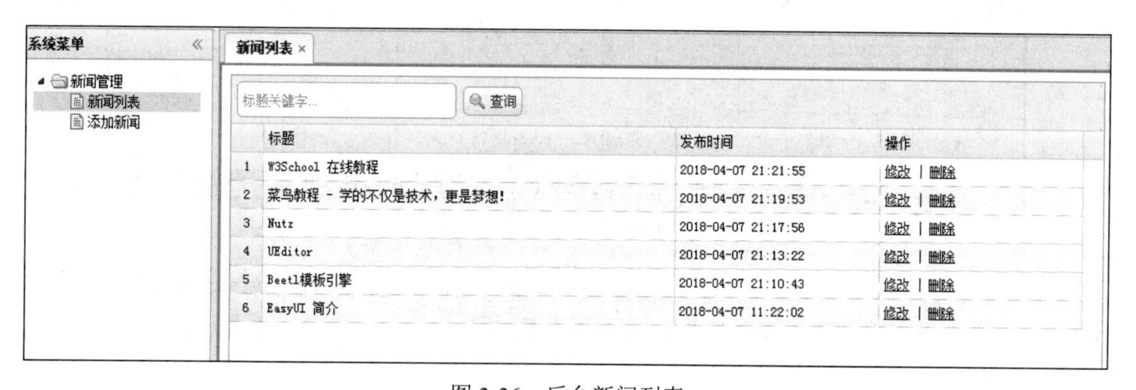

图 3-36　后台新闻列表

主要的 HTML 代码以 DataGrid 呈现新闻列表，在其附属的工具栏中显示一个关键字输入框和一个查询按钮。

图 3-37 后台查询新闻

```
<table id="dg" cellpadding="2" ></table>
<div id="tb" style="padding:5px;">
    <input id="s_name" class="easyui-textbox"data-options="prompt:'标题关键字...'" style=
"width:200px; height:32px">
    <a id="s_news" href="#" class="easyui-linkbutton" data-options="iconCls:'icon-search'">查询
</a>
</div>
```

主要 JS 代码实现 DataGrid 中新闻数据的动态分页加载，执行查询、删除和修改操作。需要注意以下几个方面：

（1）父级消息框。JS 代码用到形如 parent.$.messager 的写法，主要作用是弹出父级页面中的 EasyUI 消息框，不是当前页面（iframe）中的消息框，优势是让消息框（模态窗口）遮挡整个页面（包括父级 Layout 和当前 iframe），使得用户只能操作弹出的消息框，不能操作页面上其他内容。这种使用方法在后文还用到很多次。

（2）长标题自动折断。当标题太长时，如果在一行全部显示，会破坏页面布局效果。所以希望标题太长时，自动根据显示宽度折断，但当鼠标移动到该标题上时，显示完整的标题内容，效果如图 3-38 所示。解决方案是，对于标题的显示特殊处理，一是修改样式，实现过长自动折断，修改 easyui.css 中.datagrid-cell 样式，添加属性和值 text-overflow:ellipsis；二是增加元素的 title 属性（'<span title='+ value+ '>' + (value?value:'')+'</span>'），鼠标指针 hover 时，自动显示完整标题。

图 3-38 长标题自动折断

（3）当焦点在 id 为 tb 的 div 元素中时，按下回车键，即进行查询$("#tb").bind("keydown", **function**(e){…})；如果焦点在 tb 以外，按下回车键不会执行查询操作。代码如下：

```
var s_name="",id="",title="";
function loadGrid(){
    s_name=$("#s_name").val();
    $("#dg").datagrid({
        width:800,height:500,nowrap:false,
```

```
            striped:true,border:true,collapsible:false,
            url:"${ctxPath}/news/listNews",
             queryParams:{"s_name":s_name},
            pagination:true,
            rownumbers:true,
            fitColumns:true,pageSize:20,
            loadMsg:'数据加载中...' ,
            columns:[[
                {title:'标题', field:'title',width:200,formatter: function(value,row,index){
                    return '<span style="white-space: nowrap;" title='+value+'>'+ (value?value:'')+
'</span>';
                }},
                {title:'发布时间', field:'tjdate',width:100},
                {title:'操作', field:'hitnum',width:100, formatter: function(value,row,index){
                    var p="<a href=\"javascript:editNews('"+row.id+"')\">修改</a>";
                    p+="    |    <a href=\"javascript:delNews('"+row.id+"','"+row.title+"')\">删除</a>";
                    return p;
                    }}
                ]],
            toolbar:'#tb'
    });
}
function editNews(id){
    parent.swNewTab("修改新闻信息","${ctxPath}/news/goEdit?id="+id);
}
function delNews(newsid,title0){
    id=newsid;title=title0;
    parent.$.messager.confirm("系统提示", "您确认要删除“"+title+"”吗？", function(r){
        if (r){
            $.ajax({
                url:"${ctxPath}/news/doDel1",
                data:{"id":id},
                type:"post",
                success: function(res){
                    if(res=="true"){
                            parent.$.messager.alert("系统提示","您已删除新闻："+title, "info");
                            id="";s_name="";
                            loadGrid();
                    }else {
                            parent.$.messager.alert("系统提示",res,"error");
                    }
                    return false;
                },
                error:function(res){
                    parent.$.messager.alert("系统提示","系统错误","error");
                }
```

```
                })
            }
        });
    }

    $(function(){
        loadGrid();
        $("#s_news").click(function(){
                s_name=$("#s_name").val();
                loadGrid();
        });
            $("#tb").bind("keydown",function(e){
                var theEvent = e || window.event; //
                var code = theEvent.keyCode || theEvent.which || theEvent.charCode;
                if (code == 13) {
                    $("#s_news").click();
                }
        });
    })
```

若要设置进入后台管理页面后默认打开后台新闻列表 Tab 选项卡，只需在 admin.html 页面中添加以下 JS 代码。

```
    $(function(){
        swNewTab('新闻管理',"${ctxPath}/news/goList");
    });
```

2. 百度编辑器

添加或修改新闻时，实际上不限于纯文字的发布和修改，通常需要上传图片、视频或者其他文件，需要呈现图文混排的多样性，所以使用普通的 textarea 满足不了需求。通常采用在线编辑器，我们选用百度 UEditor 富文本 Web 编辑器，最主要的理由是：开源免费，功能丰富。

UEditor 是一套开源的在线 HTML 编辑器，主要用于让用户在网站上获得所见即所得的编辑效果。开发人员可以用 UEditor 把传统的多行文本输入框（textarea）替换为可视化的富文本输入框，涵盖流行富文本编辑器特色功能，独创多种全新编辑操作模式，屏蔽各种浏览器之间的差异。UEditor 使用 JavaScript 编写，可以无缝地与 Java、.NET、PHP、ASP 等程序集成，适合在 CMS、商城、论坛、博客、Wiki、电子邮件等互联网应用上使用。

（1）获取资源。可以从官网下载 ueditor1_4_3_3-utf8-jsp 版本完整安装包。图 3-39 所示是 ueditor1_4_3_3-utf8-jsp 版本压缩包内容。

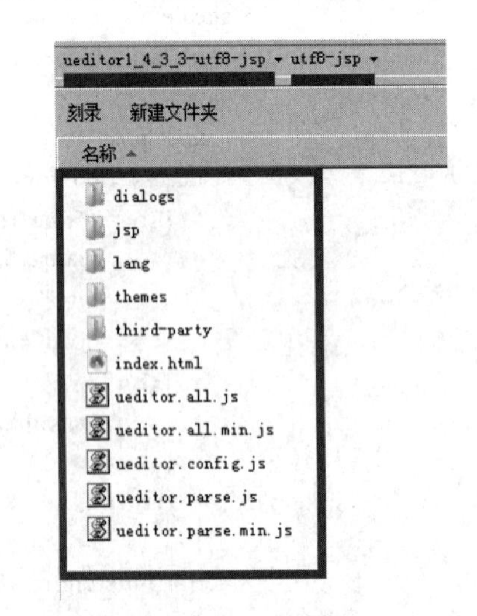

图 3-39 ueditor1_4_3_3-utf8-jsp 版本压缩包内容

（2）安装资源。

- 将下载的资源包解压缩，重命名父级文件夹，如 ueditor，放入 include 文件夹中。

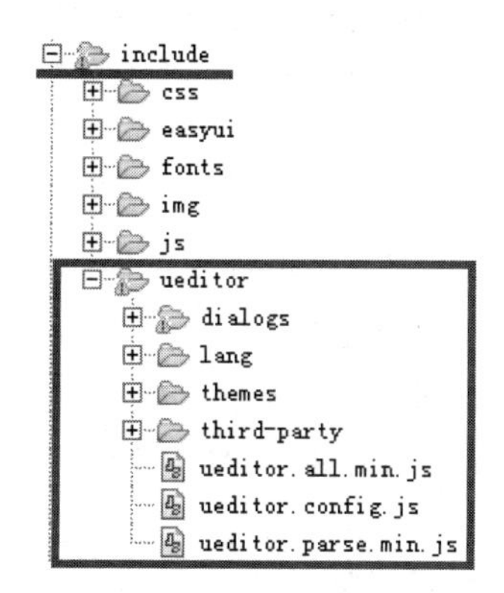

图 3-40　安装 ueditor 资源到当前项目

- 将包中 ueditor1_4_3_3-utf8-jsp\ueditor\jsp\lib 里面的*.jar 和 ueditor1.5.0.jar 放入 WEB-INF\lib 文件夹中。
- 将包中 jsp 里面的 config.json 放入 conf 文件夹中。

（3）修改配置。在 WebRoot 目录下创建文件夹 upload 作为指定存放上传文件的目录。

- 修改 ueditor.config.js 文件。指定服务器上响应文件上传的方法，其中 base 为当前项目 path。

```
serverUrl: base+"file/bdupfile",
```

- 修改 config.json 文件。把原始文件中所有默认的上传保存路径/ueditor/jsp/upload/全部修改为当前项目设计的路径前缀/upload/。例如，图片的上传保存路径修改为：

```
"imagePathFormat": "/upload/image/{yyyy}{mm}{dd}/{time}{rand:6}",
```

（4）使用样例。

```
<script>var base="${ctxPath}/";</script>
    <script type="text/javascript" charset="UTF-8" src="${ctxPath}/include/ueditor/ueditor.config.js">
</script>
    <script type="text/javascript" charset="UTF-8" src="${ctxPath}/include/ueditor/ueditor.all.min.js">
</script>
    <!--建议手动加载语言，避免在 IE 下有时因为加载语言失败导致编辑器加载失败-->
    <!--这里加载的语言文件会覆盖你在配置项目里添加的语言类型，比如你在配置项目里配置的
是英文，这里加载的中文，那最后就是中文-->
    <script  type="text/javascript"  charset="UTF-8"  src="${ctxPath}/include/ueditor/lang/zh-cn/zh-cn.js">
</script>
<div>
    <script id="editor" type="text/plain" style="width:100%;height:500px;"></script>
</div>
<script type="text/javascript">
    //实例化编辑器
    var ue = UE.getEditor('editor');
</script>
```

更多使用方法，请查看 UEditor 文档。

（5）可能存在的问题。

- 百度编辑器 Jar 包问题。WEB-INF\lib 中缺少需要的 Jar 包（commons-codec-1.9.jar、commons-fileupload-1.3.1.jar、commons-io-2.4.jar、json.jar、ueditor-xxx.jar）；或者存在重复 Jar 包，只是版本号不同；或者版本太旧。
- 配置文件的物理路径有问题。比如 tomcat 的路径上有空格，或者有汉字，运行项目后，配置文件生成到 tomcat 的 webapps 中，但由于 tomcat 的安装路径有特殊字符，导致 UEditor 不能正常读取 config.json。
- 配置文件的编码有问题。应确保 config.json 文件的编码方式 UTF-8。
- 配置文件的内容有问题。config.json 文件中不能出现“//”注释符。

3. 添加新闻

配置好 UEditor 编辑器，则可以在添加和修改新闻时，采用 UEditor 实现图片、动画、附件、视频等文件上传，实现在线图文混排。

添加新闻也是发布新闻的一个基本操作，编写本书时已经简化部分功能，忽略了发布时需要选择新闻版块、是否图片新闻、是否置顶等功能。

在 WebRoot\WEB-INF\web 中创建 newsadd.html 页面，实现新闻的前端添加发布，页面效果如图 3-41 所示。

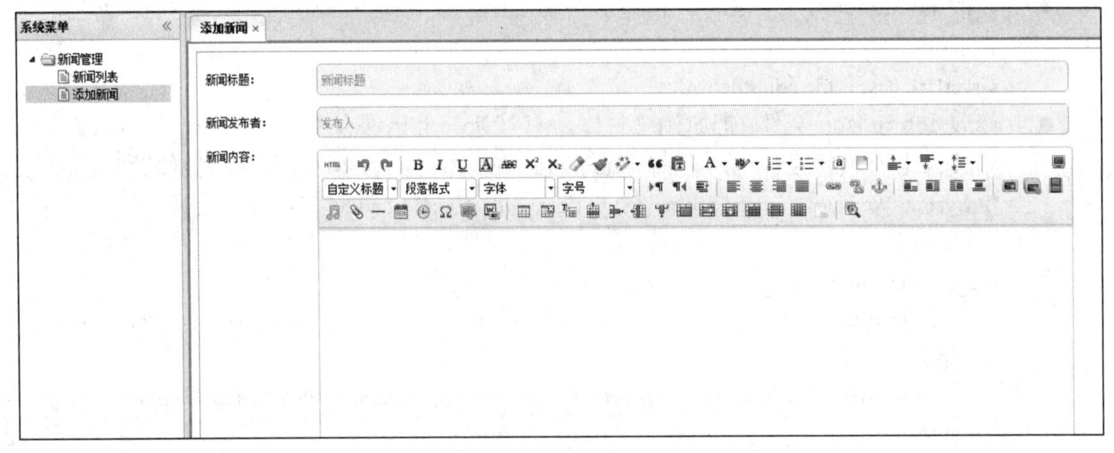

图 3-41　添加新闻窗口

（1）引用文件。

```
<link rel="stylesheet" type="text/css" href="${ctxPath}/include/easyui/themes/default/easyui.css">
<link rel="stylesheet" type="text/css" href="${ctxPath}/include/easyui/themes/icon.css">
<script src="${ctxPath}/include/js/jquery.min.js"></script>
<script type="text/javascript" src="${ctxPath}/include/easyui/jquery.easyui.min.js"></script>
<script type="text/javascript" src="${ctxPath}/include/easyui/locale/easyui-lang-zh_CN.js"> </script>
<script>var base="${ctxPath}/";</script>
<script    type="text/javascript"    charset="UTF-8"    src="${ctxPath}/include/ueditor/ueditor.config.js">
</script>
<script    type="text/javascript"    charset="UTF-8"    src="${ctxPath}/include/ueditor/ueditor.all.min.js">
</script>
<script type="text/javascript" charset="UTF-8" src="${ctxPath}/include/ueditor/lang/zh-cn/zh-cn.js"> </script>
```

（2）页面主要代码。对于新闻对象的主要属性，其中新闻 id 和新闻的发布时间由后端生成，不需要前端输入。前端仅提供新闻标题、新闻内容和新闻发布人的输入，另加一个“保存”按钮。

```
<div class="easyui-panel" style="padding:5px 2px">
<form>
    <table cellpadding="5">
    <tr><td style="width:100px;">新闻标题：</td><td style="width:880px;">
        <input id="title" class="easyui-textbox" data-options="prompt:'新闻标题', required:true"
style="width:90%;height:32px">
    </td></tr>
    <tr><td>新闻发布者：</td><td>
        <input id="cruser" class="easyui-textbox" value="${me.xm}" data-options="prompt:'发布人',
required:true" style="width:90%;height:32px">
    </td></tr>
    <tr><td style="vertical-align: top;">新闻内容：</td><td>
        <script id="content" type="text/plain" style="width:89%;height:300px;"></script>
    </td></tr>
    </table>
</form>
<div style="text-align:center;">
    <a id="savenews" href="#" class="easyui-linkbutton" iconCls="icon-ok" style="width:132px;
height:32px">保存</a>
</div>
</div>
```

（3）主要 JS 代码。实现百度编辑器的创建，新增新闻信息的校验和通过 Ajax 向后端提交。

```
var ue;
$(function(){
    ue = UE.getEditor('content');
    $('#savenews').click(function(){//发布新闻前，要校验
        var a=$("#title").textbox("getValue");
        var b=ue.getContent();
        var c=$("#cruser").textbox("getValue");
            if(a.length<=0){$.messager.alert("系统提示","必须填写新闻标题","warning"); return;}
            if(b.length<=0){$.messager.alert("系统提示","必须填写新闻内容","warning"); return;}
            if(c.length<=0){$.messager.alert("系统提示","必须填写发布人姓名或者发布机构
                名称","warning"); return;}
            $.ajax({
                type: 'POST',
                url : "${ctxPath}/news/saveAdd",
                data : {"title":a,"content":b,"cruser":c},
                success : function (res) {
                    if(res=="true"){
                            parent.$.messager.alert("系统提示","你已添加新闻:"+
                                $("#title").val(),"info");
                    }else{
                            parent.$.messager.alert("系统提示","添加失败！", "error");
                    }
                    return false;
```

```
                        },
                        error : function(res) {parent.$.messager.alert("系统提示","系统错误！", "error"); }
                    });
                });
            });
```

4. 删除新闻

在软件系统中进行“删除”操作，首先要有确认步骤，如图 3-42 所示，这在一定程度上可防止不小心误删。删除操作之后，通常会有删除成功或删除失败的提示信息。相关实现功能的 JS 代码，已在“后台新闻列表”中给出，不再赘述。

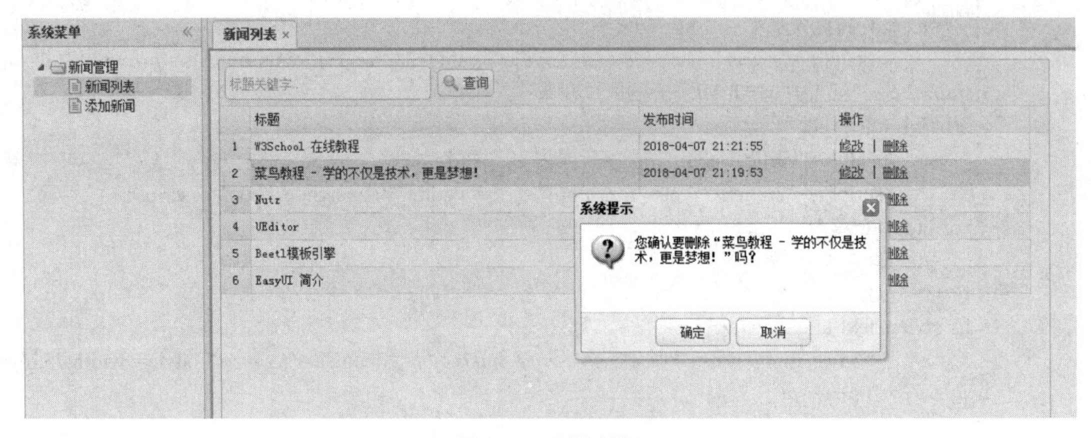

图 3-42 删除新闻

5. 修改新闻

在 WebRoot\WEB-INF\web 中创建 newsedit.html 页面，其内容与新闻添加页基本相同，如页面中最主要的部分是对应 news 对象各种属性的输入框，不相同的部分主要体现在以下几点。

（1）页面中使用${news.xxx}接收后端返回的一条新闻 news 的各种属性值，页面效果如图 3-43 所示。

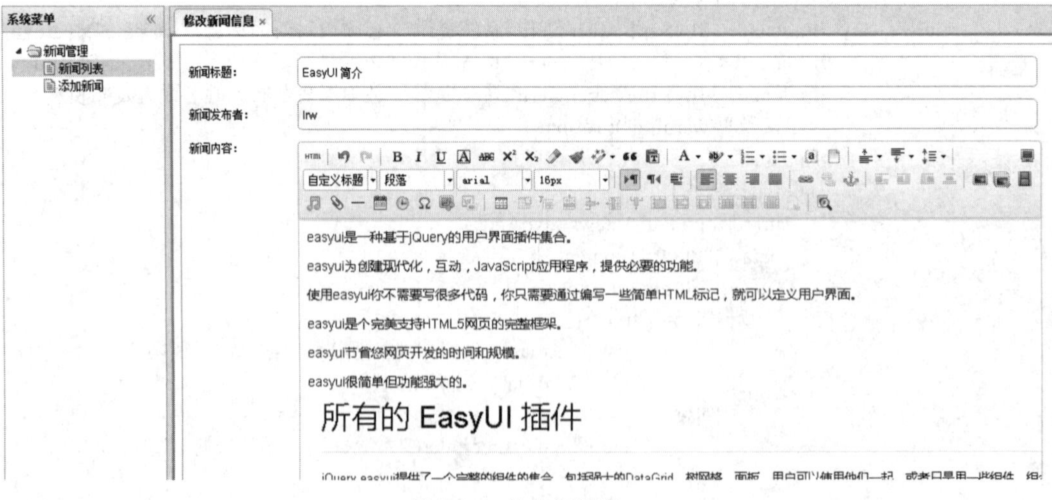

图 3-43 修改新闻窗口

（2）请求保存的路径不一样。

（3）操作结果的提示信息有区别。

下面的 JS 代码，主要实现创建百度编辑器、动态加载指定 id 新闻信息、校验后保存修改的信息。

```
var ue;
$(function(){
    ue = UE.getEditor('content');
    $("#title").textbox("setValue","${news.title}");
    ue.ready(function() {
        ue.setContent("");
        ue.execCommand('insertHtml', '${news.content}');
    });
    $("#cruser").textbox("setValue","${news.cruser}");
    $('#savenews').click(function(){//发布新闻前，要校验
        var a=$("#title").textbox("getValue");
        var b=ue.getContent();
        var c=$("#cruser").textbox("getValue");
            if(a.length<=0){$.messager.alert("系统提示","必须填写新闻标题","warning");return;}
            if(b.length<=0){$.messager.alert("系统提示","必须填写新闻内容","warning");return;}
            if(c.length<=0){$.messager.alert("系统提示","必须填写发布人姓名或者发布机构名称
","warning");return;}
            $.ajax({
                type: 'POST',
                url : "${ctxPath}/news/saveEdit",
                data : {"title":a,"content":b,"cruser":c,"id":${news.id}},
                success : function (res) {
                        if(res=="true"){
                            parent.$.messager.alert(" 系 统 提 示 "," 你 已 修 改 新 闻 :"+
$("#title").val(), "info");
                        }else{
                            parent.$.messager.alert("系统提示","修改失败！","error");
                        }
                        return false;
                },
                error : function(res) {parent.$.messager.alert("系统提示","系统错误！","error");}
            });
    });
});
```

3.8.10　前台新闻信息处理

1. 新闻列表

教学中进行简单化处理，修改系统首页 index.html，使其既有登录功能，还可以查看新闻列

表，如图 3-44 所示。仍然要处理过长的标题显示效果。新闻标题过长时可以自动截断，并在末尾显示“...”，避免太长的标题破坏页面外观效果。加载新闻列表时，通常只显示标题和发布日期。新闻条目较多时，通常采用分页控件实现分页效果，继续使用 EasyUI 的控件，如 pagination。

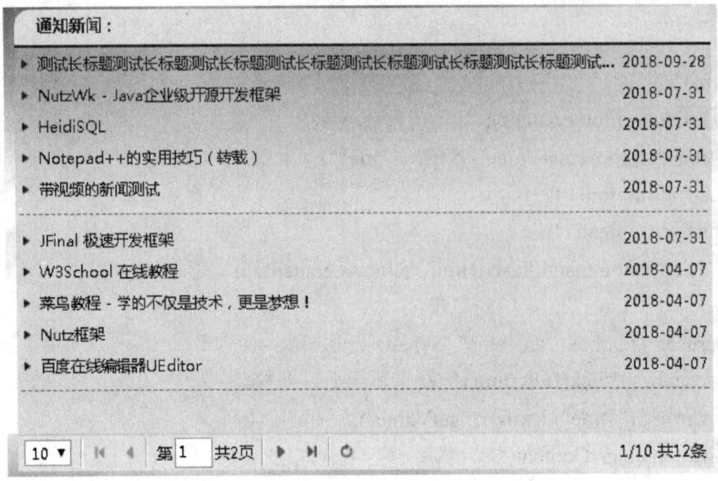

通知新闻：

▸ 测试长标题测试长标题测试长标题测试长标题测试长标题测试长标题测试长标题测试...	2018-09-28
▸ NutzWk - Java企业级开源开发框架	2018-07-31
▸ HeidiSQL	2018-07-31
▸ Notepad++的实用技巧（转载）	2018-07-31
▸ 带视频的新闻测试	2018-07-31
▸ JFinal 极速开发框架	2018-07-31
▸ W3School 在线教程	2018-04-07
▸ 菜鸟教程 - 学的不仅是技术，更是梦想！	2018-04-07
▸ Nutz框架	2018-04-07
▸ 百度在线编辑器UEditor	2018-04-07

10 ▾ 第1 共2页 1/10 共12条

图 3-44　前台新闻列表

页面仍然使用 Ajax 技术实现动态加载新闻列表。以下是 index.html 页面添加的用于显示新闻列表的主要代码，没有提供 CSS 文件。请大家发挥各自的能力，设计实现自己特色（样式）的新闻列表页。

```
<div id="lnews" class="l-wrap">
<div>
    <div>
        <div class="l-news">
            <div class="nheader">
                <table cellspacing="0" cellpadding="0"><tbody>
                    <tr><td><h3>通知新闻：</h3></td></tr>
                </tbody></table>
            </div>
            <div class="nlist">
                <table id="newstable" width="100%">
                <tbody>
                    <tr id="trpp"><td colspan="2"> </td></tr>
                </tbody></table>
            </div>
            <div id="pp" style="background:#efefef;"></div>
        </div>
    </div>
</div>
</div>
```

为了让登录窗口、新闻列表分别显示，可以在页面 body 里面加超链接，实现简单的切换。

```
<div style="float:right;padding-right:20px;">
<a id="a" href="#" style="margin-right:15px;" >登录</a>      <a id="b" href="#" >新闻</a>
</div>
```

页面中实现登录与新闻列表切换、新闻列表内容动态加载的 JS 代码如下所述。

首先通过一个 ajax，获取新闻总数（pageTotal）、加载第 1 页的新闻列表（listNews）、加载 EasyUI 分页器（loadPager）。默认隐藏登录窗口，显示新闻列表。为了能够切换登录窗口和新闻列表窗口，让超链接绑定 click 事件。

```
var pageN=1,pageTotal=100;
$(function(){
    $.ajax({
        url:"./news/getCount",
        type:"post",
        success: function(res){
            pageTotal=parseInt(res);
            listNews(1,10);loadPager();
        },
        error:function(res){
            $.messager.alert("系统提示","系统错误","error");
        }
    });
    $("#llogin").hide();
    $("#a").click(function(){
        $("#llogin").show();
        $("#lnews").hide();
    });
    $("#b").click(function(){
        $("#llogin").hide();
        $("#lnews").show();
    });
});
function listNews(pageNumber,pageSize){
    $.ajax({
        url:"./news/listNews",
        data:{"page":pageNumber,"rows":pageSize},
        type:"post",
        success: function(res){
            $(".inews").remove();
            res=JSON.parse(res);
            if(res.total<=0){
                var tr="<tr class='inews' height=\"25\"><td >";
                    tr+="<div class='t'>暂无相关新闻</div>";
                    tr+="</td><td width='1%' nowrap=''><span > </span></td></tr>";
                $("#trpp").before(tr);
```

```
                    }
                    else {
                        pageN=pageNumber;
                        pageTotal=res.total;
                        var rows=res.rows;
                        for(var i=0;i<rows.length;){
                            var row=rows[i];
                            var tr="<tr class='inews' height=\"25\"><td >";
                                tr+="<div class='t'><a href='./news/getNews?id="+row.id+"' target=
                                    '_blank' title='"+row.title+"'>"+row.title+"</a></div>";
                                tr+="</td><td width='1%' nowrap=''><span >"+row.tjdate.substr(0,10)+
                                    " </span></td></tr>";
                            $("#trpp").before(tr);
                            i++;
                            if(i%5==0){
                                tr="<tr class='inews' height='1'><td colspan='2' align='center'>";
                                tr+="<hr style='border-style:dashed;border-color:#999999;width:99%;
                                    height:1px;border-width:1px 0px 0px 0px;visibility:inherit'></td></tr>";
                                $("#trpp").before(tr);
                            }
                        }
                    }
                },
                error:function(res){
                    parent.dlg_ok(300,140,"系统提示","系统错误",null,"error");
                }
            })
        }
        function loadPager(){
            $('#pp').pagination({
                total:pageTotal,
                pageSize:10,
                pageNumber:pageN,
                displayMsg:'{from}/{to}  共{total}条',
                onSelectPage:function(pageNumber, pageSize){
                    listNews(pageNumber,pageSize);
                }
            });
        }
```

2. 新闻阅读

在 WebRoot\WEB-INF\web 中创建新闻阅读页 newsread.html，用于展示一条新闻的内容。新闻阅读页的内容通常包括：新闻标题、新闻主体内容（图、文、附件链接等）、日期、来源（发布人或机构）、阅读量，如图 3-45 所示。

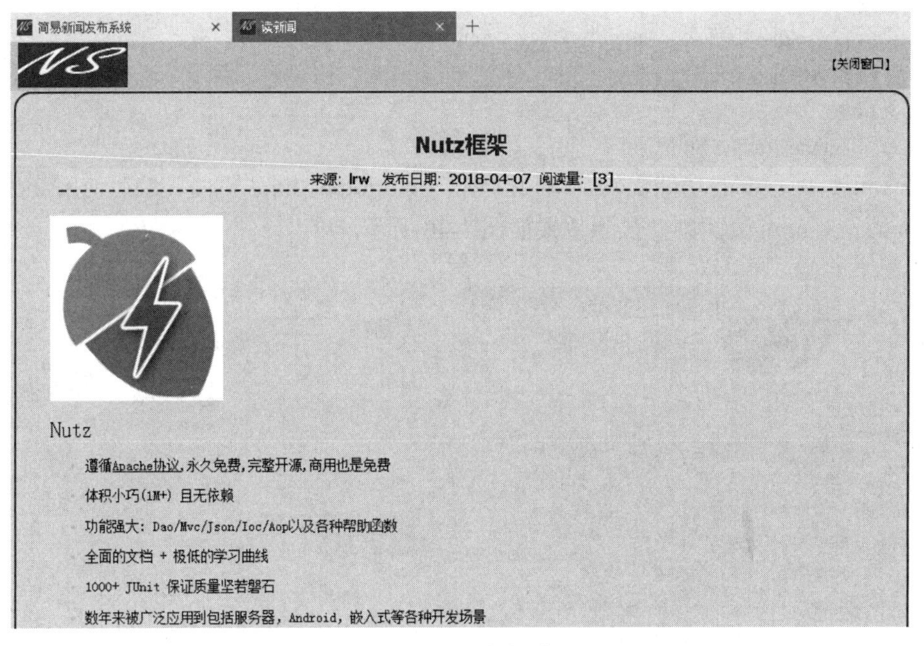

图 3-45　新闻阅读

为了防止新闻发布时间显示不正常，使用了 Beetl 方法简单转换了日期格式 ${news.tjdate,dateFormat="yyyy-MM-dd"}。

页面中增加了 2 个功能：

- 关闭窗口。实际上是关闭当前页面。
- 返回顶部。如果新闻内容很长，当向下滑动一段距离后，则自动出现“返回顶部”的功能。返回顶部的方法有很多种，比如使用 jquery.toTop.min.js 可以实现返回顶部不一样的效果。

页面主要 html 代码如下，没有提供 CSS 代码，请发挥各自的能力，实现有特色的新闻阅读页面。

```
    <div style="background:#B3DFDA;padding:0 10px 0 10px;vertical-align: middle;">
        <img src="${ctxPath}/include/img/logo.png" width="126" height="50" />
        <div style="float:right;line-height:50px;margin-right:10px;font-size: 9pt;">
            <span>【</span><a style="color:blue;" href="javascript: window.close();"><span>关闭窗口</span></a><span>】</span>
        </div>
    </div>
<div class="ndetail">
    <div class="ntitle">${news.title}</div>
    <div class="nauthor">
        <div> 来 源 :  <strong>${news.cruser}</strong>    发 布 时 间 : <strong>${news.tjdate,dateFormat="yyyy-MM-dd"}</strong>   访 问 量 :  <strong>[<span>${news.hitnum}</span>]</strong></div>
    </div>
    <div class="nbody">
```

```
            <div id="vsb_content"> ${news.content}</div>
        </div>
    </div>
    <% include("totop1.html"){}%>
```

以下实现返回顶部的功能代码，单独创建了一个 totop1.html 页面，然后使用 include 将其包含到 newsread.html 页面中，显示效果如图 3-46 所示。

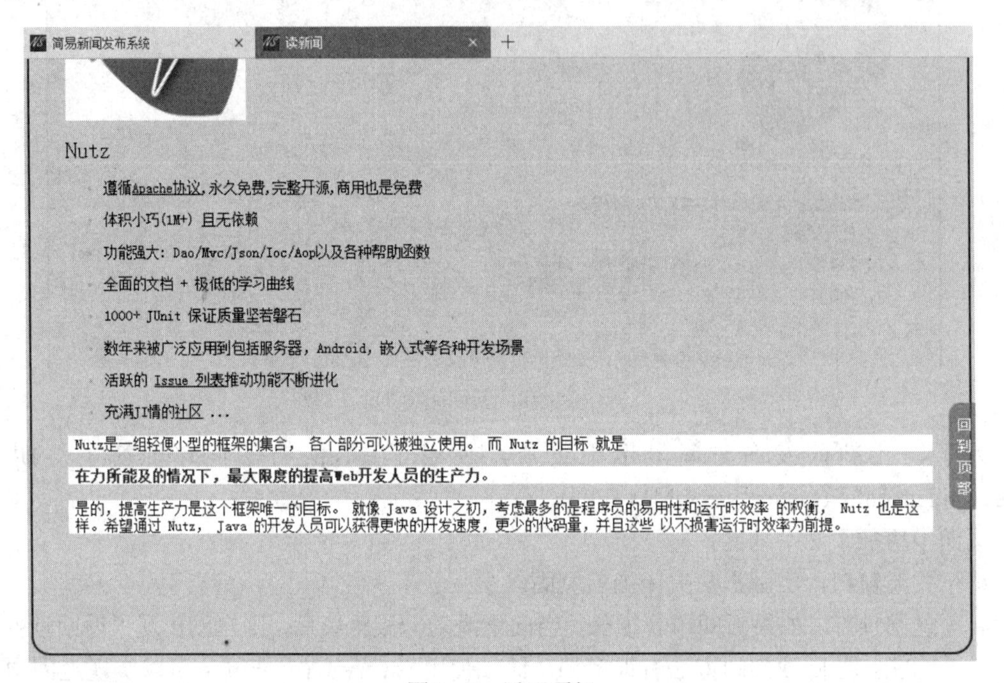

图 3-46　返回顶部

```
<style>
#backtotop {
    width: 24px;
    color: white;
    padding: 12px 0px 12px 5px;
    display: none;
    position: fixed;
    cursor: pointer;
    text-align: center;
    z-index: 20;
    background-color: rgba(0, 188, 212, 0.65);
    border-radius: 12px 0px 0px 12px;
}
</style>
<div id="backtotop" style="right: 0px; display: none;">回到顶部</div>
<script>
var $backtotop = $('#backtotop');
```

```
var toplrw = $(window).height() - $backtotop.height() - 200;

function moveBacktotop() {
    if(toplrw<0) toplrw=0;
    $backtotop.css({ top: toplrw, right: 0});
}

$backtotop.click(function () {
    $('html,body').animate({ scrollTop: 0 });
    return false;
});
$(window).scroll(function () {
    var windowHeight = $(window).scrollTop();
    if (windowHeight > 200) {
    $backtotop.fadeIn();
    } else {
    $backtotop.fadeOut();
    }
});

moveBacktotop();
$(window).resize(moveBacktotop);
</script>
```

3.9　考核任务

正确实现新闻信息的增删改查，添加与修改新闻可以上传图片等文件。

（1）添加新闻。（20 分）

（2）修改新闻。（20 分）

（3）后台新闻列表。（20 分）

（4）删除新闻。（20 分）

（5）前台新闻列表。（10 分）

（6）阅读新闻。（10 分）

本章小结

本章以快速迭代的方法实现项目，后端采用 Nutz 框架、前端采用 EasyUI 框架，详细地介绍了新闻发布系统的开发过程，循序渐进。从搭建框架项目到添加数据源和主模块、添加日志系统、添加处理业务逻辑的子模块、添加文件上传功能都有详细介绍；包含后端开发（Java 类文件、配置文件）和前端开发（前端页面 HTML 编码、CSS 样式设置、JS 前端业务逻辑）；

前端请求通过 Nutz 框架的@At 找到后端相应的 Java 方法响应。基本上每完成一个功能，就可以运行测试和使用。

本章强调软件开发人员熟练掌握后端和前端的调试方法，程序调试是分析 BUG 和解决问题最直接的能力。没有人能够保证软件写出来不用调试就可以运行正常，也没有人可以保证软件永远不会出 BUG。所以，熟练使用调试器是一个软件开发人员需要具备的基本技能。掌握调试方法，及时改正软件或程序中发现的 BUG，有助于软件的正确开发，有助于软件的后续功能开发，有助于提高软件开发人员的兴趣、提升软件开发的能力。

第 4 章　基于 SSH 的项目实战

本章介绍采用前端和后端分离开发的方式来开发基于 SSH 的项目。首先采用向导搭建 SSH 框架的 Web 项目（主要是引入 SSH 三大框架的 jar 包，自动配置一些参数）；然后编写后端的 Java 类（bean、dao、service、Action），配置 Spring、Struts、Hibernate 等；最后参照第 3 章介绍的前端页面内容快速开发当前项目的前端页面。

4.1　SSH 简介

SSH 是目前较流行的一种 Web 应用程序开源框架，它是 Struts+Spring+Hibernate 的一个集成框架，如图 4-1 所示。

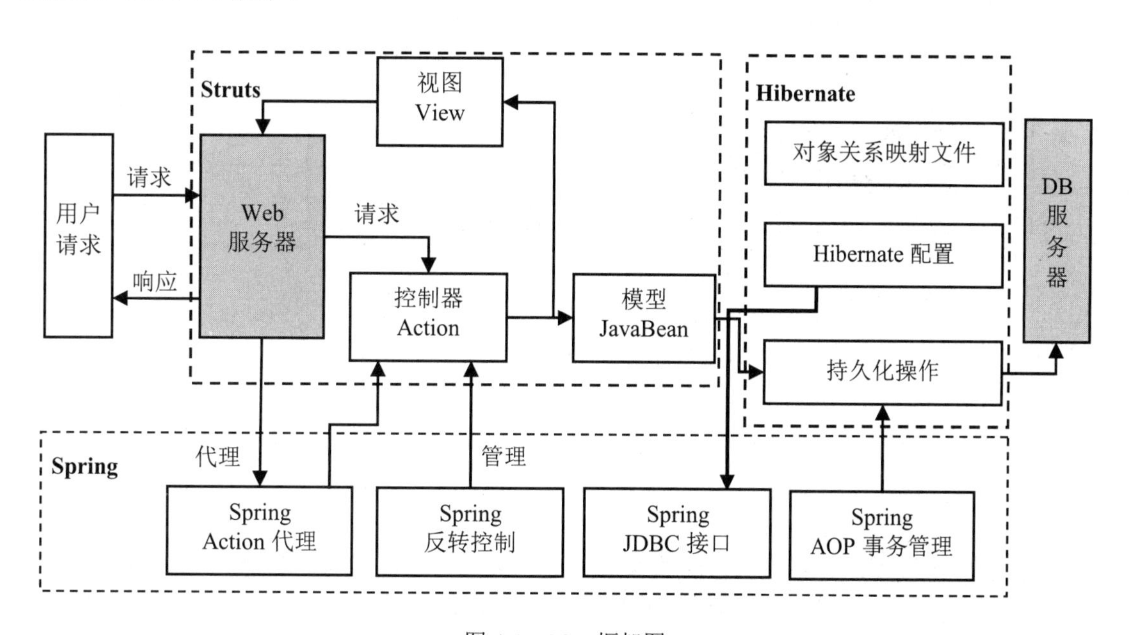

图 4-1　SSH 框架图

SSH 框架最主要的本质是“高内聚、低耦合”。其中 Struts 作为系统的整体基础架构，负责 MVC（Model、View 和 Controller）的分离，控制业务跳转。MVC 设计模式可以使系统逻辑变得很清晰。Spring 的 IOC 和 AOP 可以使产品模块在最大限度上解藕，管理 Struts 和 Hibernate。Hibernate 是实体对象的持久化，利用 Hibernate 框架对持久层提供支持，Hibernate 的 DAO（Data Access Objects）实现 Java 类与数据库之间的转换和访问。

SSH 集成框架系统包括四层，如图 4-2 所示，它可以帮助开发人员在短期内搭建结构清晰、可复用性好、维护方便的 Web 应用程序。

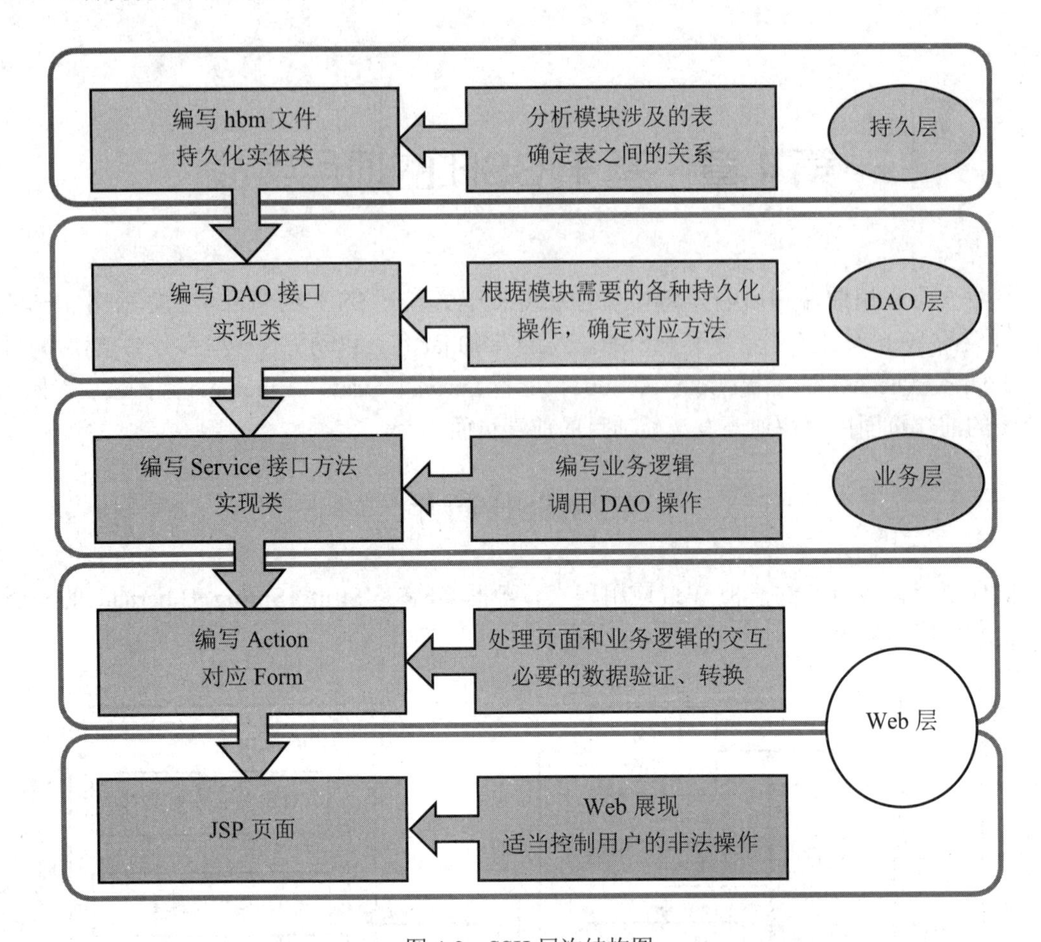

图 4-2　SSH 层次结构图

基于 SSH 的项目系统的基本业务流程如下：

- 在 Web 层中，首先通过前端 JSP 页面实现用户与系统的交互，JSP 页面简单控制用户的非法操作；然后接收用户请求，也可把后端传过来的响应，呈现到 JSP 页面；最后 Struts 根据 Struts 配置文件（struts.xml）将 ActionServlet 接收到的 Request 委派给相应的 Action 处理。
- 在业务层中，管理服务组件的 Spring IoC 容器（控制反转）负责向 Action 提供业务模型组件、调用相应 DAO 操作完成业务逻辑，其间，Spring 提供事务处理、缓冲池等，以提升系统性能和保证数据的完整性。
- 在持久层中，依赖于 Hibernate 的对象化映射和数据库交互，处理 DAO 请求的数据，并返回处理结果。

基于 SSH 的项目业务流程不仅实现了视图 View、控制器 Controller 与模型 Model 的彻底分离，而且还实现了业务逻辑层与持久层的分离。因此，前端发生变化时，模型层只需很少的改动，数据库的变化对前端的影响也会很小。这样大大提高了系统的可复用性，而且由于层之间耦合度小，有利于团队成员并行工作，大大提高了开发效率。

4.2　向导式创建 SSH 项目

使用向导方式创建 SSH 项目，MyEclipse 提供的工具可以帮助快速添加 SSH 框架需要的 Jar 包，另外该工具还可自动生成一些配置文件。

4.2.1　项目工程结构

开发框架软件项目要重视工程结构的设计，当前项目的工程结构设计见表 4-1。一些类的编写，合并了接口 Interface 与实现 Impl 类。

表 4-1　项目工程结构

路径	说明
newsssh	项目名称
├─src	源码文件夹
│　├─cn.lrw.newsssh.action	存放控制器类（合并了接口 Interface 与实现 Impl 类）
│　├─cn.lrw.newsssh.bean	存放数据库表/视图对应的 Bean 类、hbm
│　├─cn.lrw.newsssh.dao	DAO 类，封装对数据访问的一些方法（合并了接口 Interface 与实现 Impl 类）
│　├─cn.lrw.newsssh.service	业务逻辑类（合并了接口 Interface 与实现 Impl 类）
│　├─cn.lrw.newsssh.utils	公用方法类，如密码加密方法、JSON 转换等
├─config	
│　├─applicationContext.xml	Spring+Hibernate 配置，如数据源、事务管理等
│　├─config.json	系统日志配置
│　├─log4j.properties	系统日志配置
│　└─struts.xml	Struts 配置，建立前端请求与后端控制类中响应方法的联系
└─WebRoot	Web 项目根目录
├─error	存放异常访问提示页，如 nologin.jsp、403.jsp、404.jsp 等
├─include	分类存放网页中引用的 JS、CSS、图像类文件
│　├─css	
│　├─js	
│　├─img	
│　├─easyui	EasyUI 框架
│　└─ueditor	百度的可视化 HTML 编辑器
├─tmp	百度上传的临时文件夹
├─upload	存放上传的文件
├─WEB-INF	Java 的 Web 应用的安全目录，自动存放 class 文件
│　├─lib	存放项目需要的 Jar 包
│　├─web	存放网站中网页文件，如 Jsp 等
│　└─web.xml	Web 工程的配置文件
└─index.jsp	Web 工程默认的首页文件

4.2.2 准备 Jar 包和 JS 库

本项目使用表 4-2 所示的 Jar 包，还有更多需要的 Jar 包在通过向导添加 Struts、Spring、Hibernate 时自动添加，如果需要其他的 Jar 包，可以从 Maven 查找和下载。

表 4-2 项目所需 Jar 包

Jar 名	说明
mysql-connector-java	MySQL 数据库驱动
Log4j	用 Java 编写的可靠、快速和灵活的日志框架
Gson	google 开发的 Java API，用于转换 Java 对象和 JSON 对象
shiro	用于认证、授权、加密、会话管理、与 Web 集成、缓存等
UEditor	百度在线编辑器。依赖 json.jar、commons-codec-1.10.jar、commons-io-2.4.jar、commons-fileupload-1.3.3.jar、commons-lang3-3.2.1.jar
SSH 框架需要的 Jar	通过 MyEclipse 向导添加，然后作简单的清理

本项目还将使用到一些常用的 JS 库文件，如 jQuery，EasyUI，UEditor 等。

4.2.3 新建 Web 项目

（1）新建 Web Project，设置基本参数如图 4-3 所示。

- Project name：newsssh。
- Java EE version：JavaEE 7 – Web 3.1。
- Java version：1.8。
- JSTL Version：1.2.2。
- Target runtime：Tomcat v8.0。

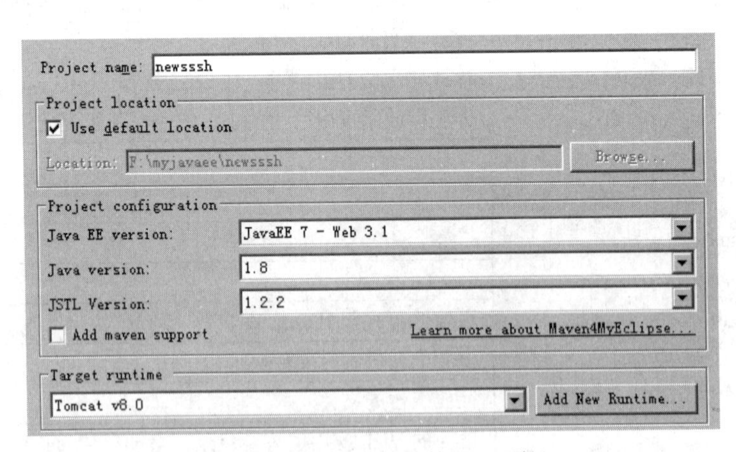

图 4-3 创建基于 SSH 的 Web 项目

（2）单击 Next，在弹出的窗口中添加 config 文件夹，用于存放后端配置文件，如图 4-4 所示。

（3）最后，勾选生成 index.jsp 和 web.xml，如图 4-5 所示。

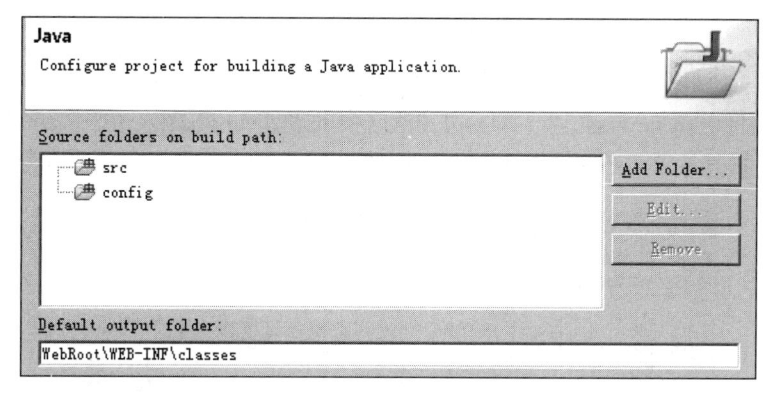

图 4-4　添加 config 文件夹

图 4-5　勾选生成 index.jsp 和 web.xml

（4）把已经准备好的 Jar 包放进项目的 WEB-INF/lib 中，如图 4-6 所示。运行测试，确保控制台没有报错信息，而且在浏览器可以正确显示首页。

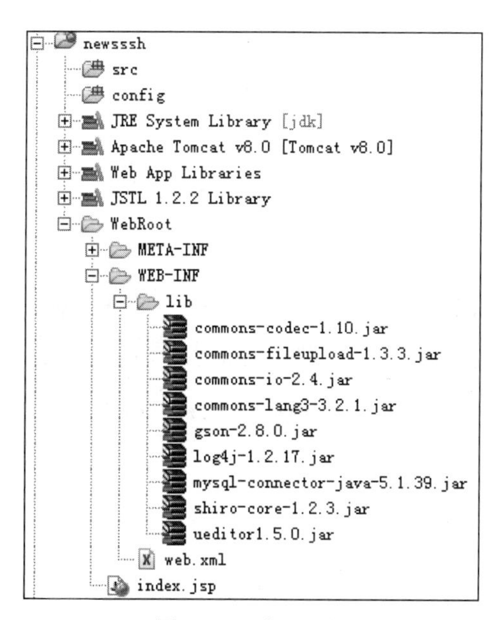

图 4-6　添加 Jar 包

（5）参照“项目工程结构”创建 package 包或者文件夹。

4.2.4 添加 Struts

右键单击项目名称 newsssh，选择 MyEclipse 中的 Project Facets，单击 Install Apache Struts（2.x） Facet，打开安装 Struts 的窗口，如图 4-7 所示，选择可以支持的 2.1 版本。

图 4-7 选择 Struts 版本

当一个请求发送到 servlet 容器的时候，容器先会将请求的 URL 减去当前应用上下文的路径作为 servlet 的映射 URL。比如访问的是 http://localhost/test/aaa.action，当前应用上下文是 test，容器会将 http://localhost/test 去掉，剩下的/aaa.action 部分作为 servlet 的映射匹配。映射匹配过程有先后顺序，当有一个 servlet 匹配成功后，其他的 servlet 将被忽略。其匹配规则和顺序如下：

（1）精确路径匹配，如<url-pattern>/test/list.do</url-pattern>，只会访问 list.do 这个 servlet。

（2）扩展名匹配，如果 URL 最后一段包含扩展名，容器将会根据扩展名选择合适的 servlet，如 servletA 的 url-pattern:*.action。

（3）路径匹配，如 servletA 的 url-pattern 为/test/*，此时访问 http://localhost/test/a 时，容器会选择路径 test 下的同名 servlet 来匹配。

如果不想在请求时带扩展名，可以选择 URL pattern 为/*，如图 4-8 所示，后续其他内容默认选择。完成安装后，在 src 中生成了 struts.xml，为了便于管理，可将它移到 config 文件夹下。

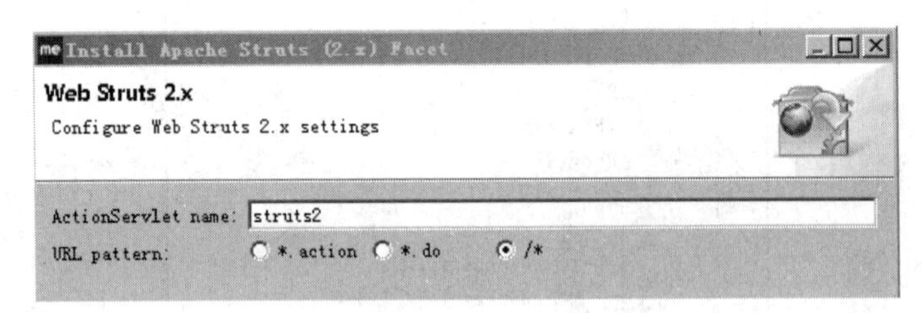

图 4-8 选择 URL 映射模式

4.2.5 添加 Spring

右键项目名称 newsssh，选择 MyEclipse 的 Project Facets，单击 Install Spring Facet，在弹出的窗口中选择 Spring 版本为 4.1，如图 4-9 所示。单击 Next，在弹出的新窗口中选择由“MyEclipse Library”配置 Spring 库（Configure Spring Libraries），配置文件名为 applicationContext.xml，存放路径更改到工程结构中的 config 文件夹，其他选项默认，如图 4-10 所示。

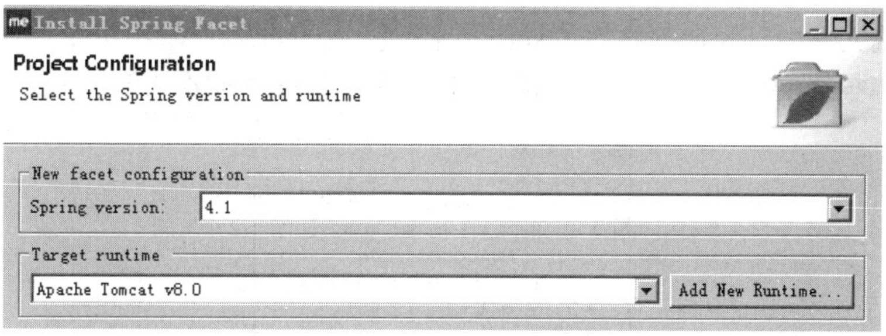

图 4-9　选择 Spring 版本

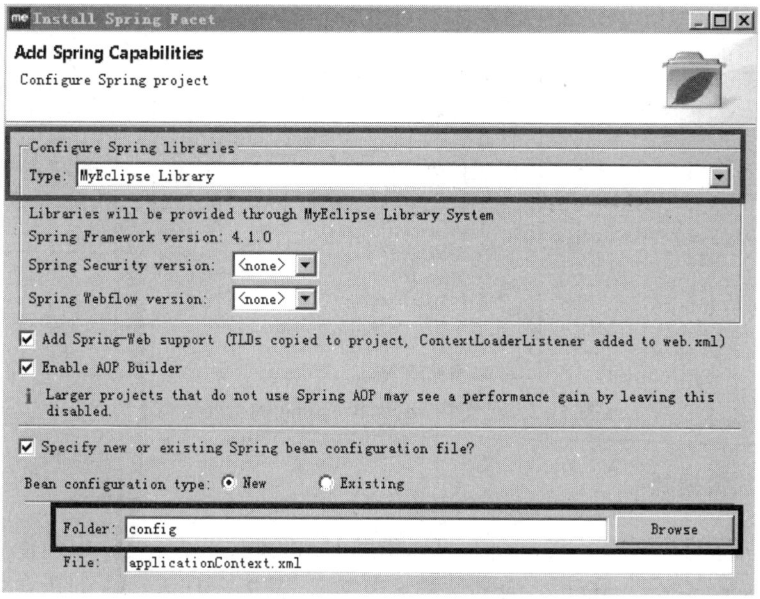

图 4-10　配置 Spring

在下一步的窗口中，选择需要安装的 Spring 库，勾选 Core、Facets、Spring Persistence 和 Web，如图 4-11 所示。

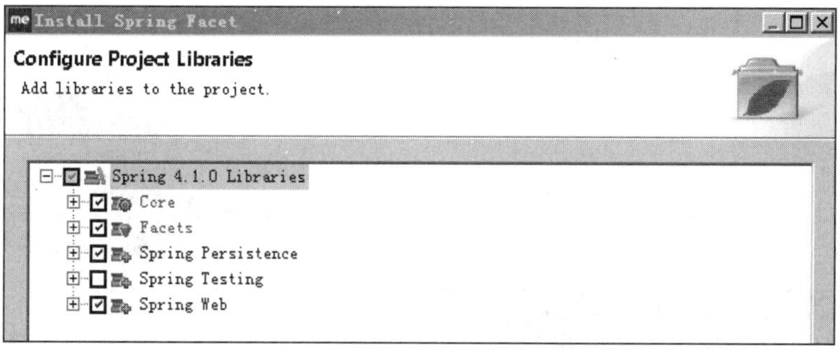

图 4-11　选择需要安装的 Spring 库

完成 Spring 的安装后，项目工程结构如图 4-12 所示，在 config 文件夹下有了两个很重要的配置文件，一个是 Spring 的配置文件 applicationContext.xml，另一个是 Struts 的配置文件

struts.xml；配置文件中已经自动添加了 spring 监听器 listener；在 WEB-INF 目录中增加了两个 Spring 的标签库描述文件（tld），用于指定关于标签处理程序的类名和标签允许的属性。

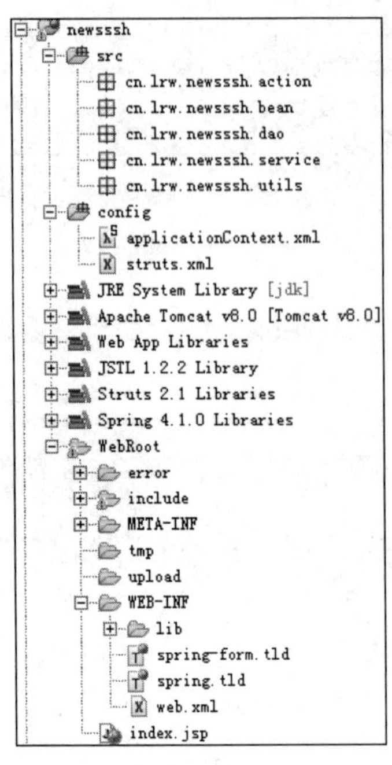

图 4-12　已安装 Struts 和 Spring 的工程结构

4.2.6　添加数据源

假定项目所需的数据库是 dbnews2，可以借助 HeidiSQL 在 MySQL 中创建数据库 dbnews2 及其 3 张表：user、news、cmenu，表的结构见表 2-1、表 2-2 和表 2-3。创建完成后，在 cmenu 表中，输入菜单数据，如图 4-13 所示。

在 MyEclipse 中，选择 Window 菜单下的 Open Perspective 命令，单击 Myeclipse Database Explorer，打开数据库浏览器，在数据库浏览器的空白区域单击右键，在弹出的快捷菜单中可以选择新建（New）连接、打开（Open）连接、编辑（Edit）连接和删除（Delete）连接命令，如图 4-14 所示。

dbnews.cmenu: 3 rows total (approximately)

id	pid	name	url	permission
1	0	新闻管理	(NULL)	1
2	1	新闻列表	goListNews	1
3	1	添加新闻	goAddNews	1

图 4-13　菜单表 cmenu 数据

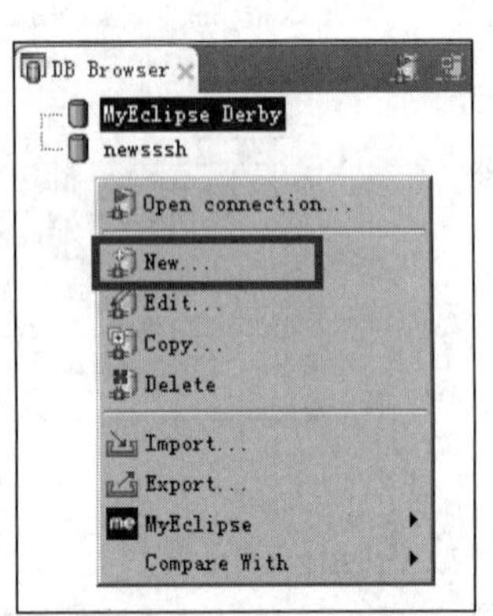

图 4-14　数据库浏览器

创建（New）与数据库 dbnews2 的连接，设定连接名称为 newsssh，如图 4-15 所示。可根据实际情况选择或设置正确的参数。在此处创建连接的目的，一是安装 Hibernate 时需要；二是在创建 bean 类时，将使用 MyEclipse 提供的逆向工程提高开发效率。参数设置如下：

- Driver：MySQL Connector/J。
- Driver name：newsssh。
- Connection URL：jdbc:mysql://localhost:3066/dbnews2?serverTimezone=Hongkong。
- User name：root。
- Password：填写连接数据库 dbnews2 的密码，如果没有密码，则保留为空。
- Driver JARs：当前项目 lib 中的 mysql-connector-java-5.xxx.jar，不宜使用最新的驱动包。
- Driver classname：com.mysql.jdbc.Driver。
- 勾选 Save Password，避免每次连接都要输入密码。

当上述参数都已经设置完成，单击 TestDriver 按钮测试连接，如果弹窗提示 Database connection successfully established.，则表明设置正确；否则查找失败的原因，如 MySQL 是否启动，数据库和表是否创建完成，参数设置是否正确等。

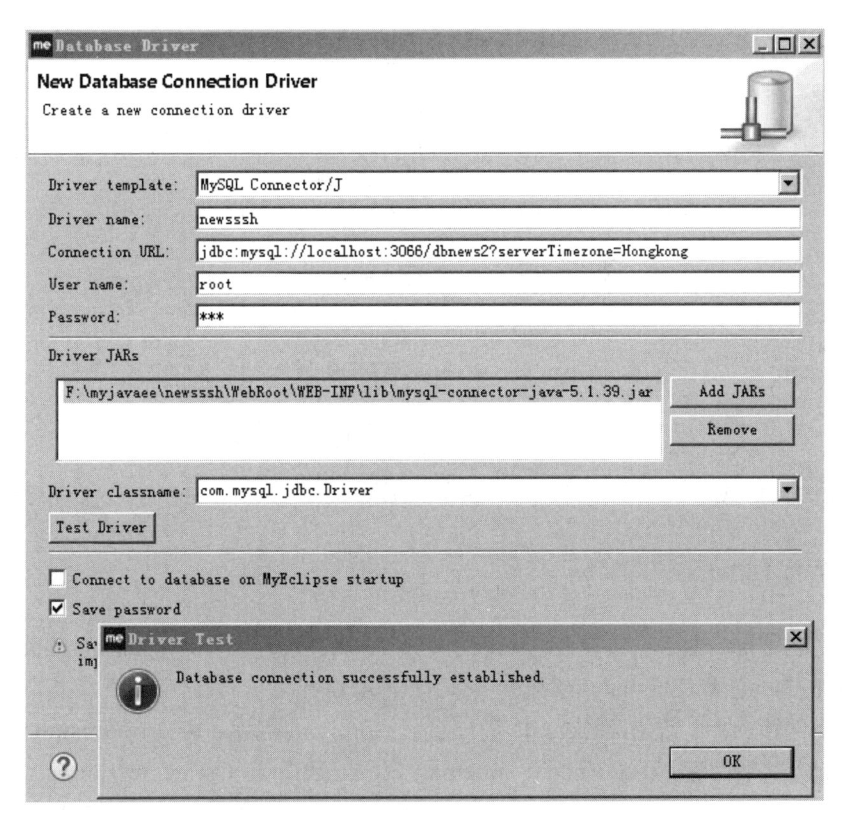

图 4-15　创建与数据库 dbnews2 的连接

测试成功后选择当前 MySQL 中的数据库，如图 4-16 所示，勾选 Display the selected schemes，然后单击 Add 按钮，选择当前项目使用的数据库 dbnews2。

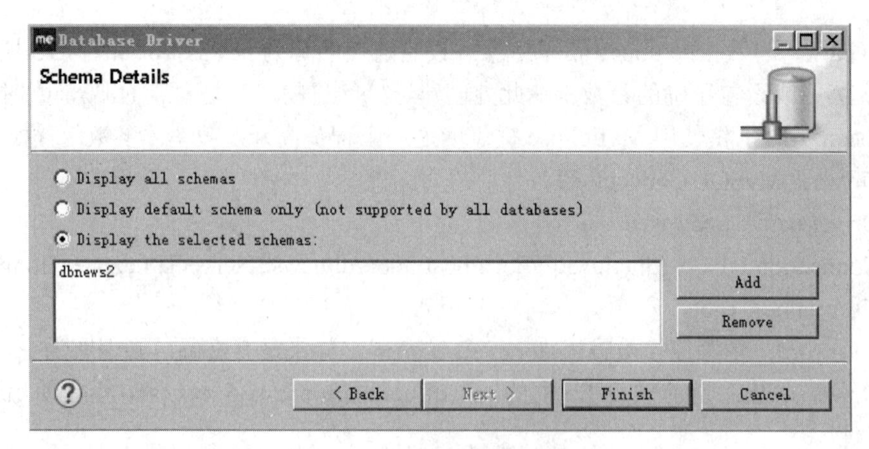

图 4-16 选择数据库 dbnews2

完成 newsssh 连接的创建后，打开连接（Open Connection），则可以看到 dbnews2 中的表，如果有视图、触发器等，都可以查看和编辑，如图 4-17 所示。

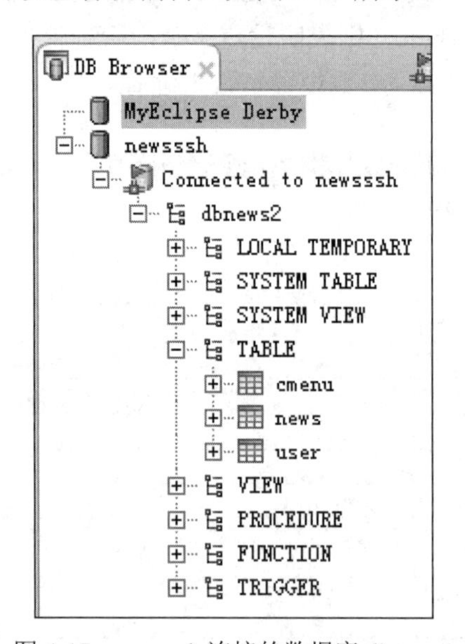

图 4-17 newsssh 连接的数据库 dbnews2

4.2.7 添加 Hibernate

右键项目名称 newsssh，选择 MyEclipse 中的 Project Facets，单击 Install Hibernate Facet，在弹出的窗口中，选择 Hibernate 的版本 4.1，如图 4-18 所示。

在下一步的窗口中，设置 Hibernate 的配置和 Spring 的配置，因为同在 applicationContext.xml 方位中，所以不勾选 Create/specify hibernate.cfg.xml file 和 Create SessionFactory class，如图 4-19 所示。

接下来配置 Spring 的 DataSource，如图 4-20 所示。设置 DB Driver 时，选择上一节中创建的数据源连接 newsssh，其他参数将会自动添加进来。不要勾选 copy DB Driver jar(s) to project and add to buildpath，因为上一节已经选择了当前项目 lib 中的 MySQL 的驱动连接 Jar 包，后续其他选项默认，直到安装完成。

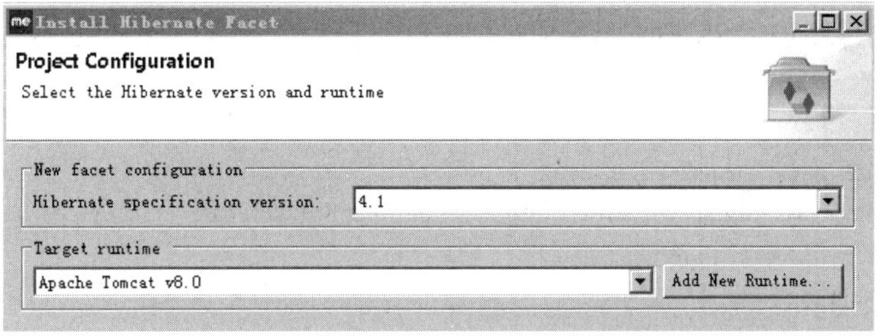

Install Hibernate Facet

Project Configuration

Select the Hibernate version and runtime

New facet configuration

Hibernate specification version: 4.1

Target runtime

Apache Tomcat v8.0　Add New Runtime...

图 4-18　选择 Hibernate 的版本

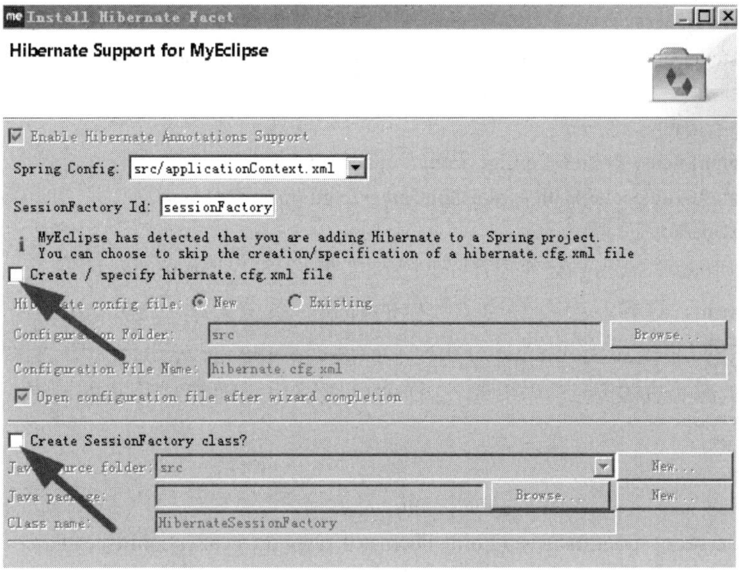

Install Hibernate Facet

Hibernate Support for MyEclipse

Enable Hibernate Annotations Support

Spring Config: src/applicationContext.xml

SessionFactory Id: sessionFactory

MyEclipse has detected that you are adding Hibernate to a Spring project.
You can choose to skip the creation/specification of a hibernate.cfg.xml file

Create / specify hibernate.cfg.xml file

Hibernate config file: New　Existing

Configuration Folder: src　Browse...

Configuration File Name: hibernate.cfg.xml

Open configuration file after wizard completion

Create SessionFactory class?

Java source folder: src　New...

Java package:　Browse...　New...

Class name: HibernateSessionFactory

图 4-19　配置 Hibernate

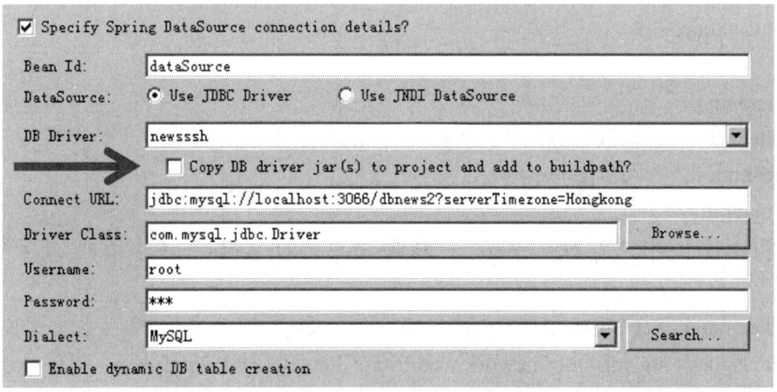

Specify Spring DataSource connection details?

Bean Id: dataSource

DataSource: Use JDBC Driver　Use JNDI DataSource

DB Driver: newsssh

Copy DB driver jar(s) to project and add to buildpath?

Connect URL: jdbc:mysql://localhost:3066/dbnews2?serverTimezone=Hongkong

Driver Class: com.mysql.jdbc.Driver　Browse...

Username: root

Password: ***

Dialect: MySQL　Search...

Enable dynamic DB table creation

图 4-20　配置 Hibernate

再次运行项目测试，确保控制台没有报错信息，而且在浏览器能正确显示首页。

4.2.8 配置 web.xml

（1）通过向导添加了 SSH 框架，已在 web.xml 里自动配置了项目对 Struts2 的支持和 Spring 监听器。

```
<filter>
    <filter-name>struts2</filter-name>
    <filter-class>org.apache.struts2.dispatcher.ng.filter.StrutsPrepareAndExecuteFilter</filter-class>
</filter>
<filter-mapping>
    <filter-name>struts2</filter-name>
    <url-pattern>/*</url-pattern>
</filter-mapping>
<listener>
    <listener-class>org.springframework.web.context.ContextLoaderListener</listener-class>
</listener>
<context-param>
    <param-name>contextConfigLocation</param-name>
    <param-value>classpath:applicationContext.xml</param-value>
</context-param>
```

（2）当 Hibernate+Spring 配合使用的时候，如果设置了 lazy=true，读取父数据后，Hibernate 会自动关闭 Session，这样，当要使用子数据的时候，系统会抛出 lazyinit 的错误，这时就需要使用 Spring 提供的 OpenSessionInViewFilter。OpenSessionInViewFilter 主要是保持 Session 状态，直到 request 将全部页面发送到客户端，这样就可以解决延迟加载带来的问题，所以在 web.xml 中添加以下配置。

```
<filter>
    <filter-name>openSessionInView</filter-name>
<filter-class>org.springframework.orm.hibernate4.support.OpenSessionInViewFilter</filter-class>
    <init-param>
      <param-name>singleSession</param-name>
      <param-value>true</param-value>
    </init-param>
</filter>
<filter-mapping>
    <filter-name>openSessionInView</filter-name>
    <url-pattern>/*</url-pattern>
</filter-mapping>
```

（3）为了防止出现中文乱码，需统一设置编码，在 web.xml 中添加以下配置。

```
<filter>
    <filter-name>Set Character Encoding</filter-name>
    <filter-class>org.springframework.web.filter.CharacterEncodingFilter</filter-class>
    <init-param>
        <param-name>encoding</param-name>
        <param-value>UTF-8</param-value>
    </init-param>
```

```
</filter>
<filter-mapping>
    <filter-name>Set Character Encoding</filter-name>
    <url-pattern>/*</url-pattern>
</filter-mapping>
```

4.2.9 配置 Spring

通过导向添加 SSH 框架，在配置文件 applicationContext.xml 中可以看到已经配置了数据源 dataSource、会话工厂 sessionFactory 和事务管理 transactionManager，在 dataSource 的 url 属性之前添加 MySQL 驱动：

```
<property name="driverClassName" value="com.mysql.jdbc.Driver"></property>
```

到目前为止，applicationContext.xml 文件的完整内容如下，手工只添加了一行。

```
<?xml version="1.0" encoding="UTF-8"?>
<beans
    xmlns="http://www.springframework.org/schema/beans"
    xmlns:xsi="http://www.w3.org/2001/XMLSchema-instance"
    xmlns:p="http://www.springframework.org/schema/p"
    xsi:schemaLocation="http://www.springframework.org/schema/beans
http://www.springframework.org/schema/beans/spring-beans-4.1.xsd
http://www.springframework.org/schema/tx    http://www.springframework.org/schema/tx/spring-tx.xsd"
xmlns:tx="http://www.springframework.org/schema/tx">

    <bean id="dataSource" class="org.apache.commons.dbcp.BasicDataSource">
        <property name="driverClassName" value="com.mysql.jdbc.Driver"></property>
        <property name="url"
            value="jdbc:mysql://localhost:3066/dbnews2?serverTimezone=Hongkong">
        </property>
        <property name="username" value="root"></property>
        <property name="password" value="123"></property>
    </bean>
    <bean id="sessionFactory"
        class="org.springframework.orm.hibernate4.LocalSessionFactoryBean">
        <property name="dataSource">
            <ref bean="dataSource" />
        </property>
        <property name="hibernateProperties">
            <props>
                <prop key="hibernate.dialect">
                    org.hibernate.dialect.MySQLDialect
                </prop>
            </props>
        </property>
    </bean>
    <bean id="transactionManager"
        class="org.springframework.orm.hibernate4.HibernateTransactionManager">
        <property name="sessionFactory" ref="sessionFactory" />
    </bean>
    <tx:annotation-driven transaction-manager="transactionManager" /></beans>
```

4.2.10 运行项目

运行基于 SSH 框架的 JavaEE 项目，如果控制台 Console 没有报错信息，则在浏览器可以看到正常的首页信息。

4.2.11 清理 Jar 包

通过向导添加 SSH 框架，可能会存在以下问题：

（1）Spring 和 Hibernate 的版本不兼容，导致控制台报错，提示 openSession 方法不存在。

（2）Struts 里面的 antlr-2.7.2.jar 和 Hibernate 中的 anltr-2.7.7.jar 冲突，造成 hql 无法执行。

（3）安装 Hibernate 不正确，需要重新安装，方法如下：

1）右键项目名称，单击 Properties 中的 Java Build Path，在右边窗口中切换到 Libraries 选项卡，选中已安装的 Hibernate 库，然后执行删除（Remove）命令。

2）删除 applicationContext.xml 中与 Hibernate 相关的配置。

3）关闭 MyEclipse。

4）打开磁盘上当前项目的目录，删除.myhibernatedata 文件。

5）打开.project 文件，删除与 Hibernate 相关的标签对。

6）打开.settings 目录下的 org.eclipse.wst.common.project.facet.core.xml 文件，删除 facet="me.hibernate"所在的行。

7）打开 MyEclipse，重新添加 Hibernate。

企业里面通常会要求创建 maven 项目。maven 项目的好处之一是在 pom.xml 中指定了需要的 Jar 包，不存在 Jar 包冲突，但需要连接到 maven 仓库。

当前项目比较简单，统一使用本地的 Jar 包。使用向导添加 SSH 三个框架时，存在引用了同名但不同版本号的 Jar 包，这样会影响后续的操作。这里提供一种解决方案：先清理，重新引用。具体方法如下：

- 添加 SSH 框架后，在运行测试正常的前提下，停止 Tomcat 服务器的运行。
- 打开 tomcat 服务器 webapps 中的 newsssh 项目。
- 清理 newsssh\WEB-INF\lib 中的 Jar 包，一般规则是：删除同名但版本号较低的 Jar 包。
- Struts2 必需的 Jar 包：

  commons-fileupload-1.3.1.jar
  commons-io-2.4.jar
  commons-lang3-3.2.1.jar
  dom4j-1.6.1.jar
  freemarker-2.3.16.jar
  ognl-3.0.jar
  struts2-core-2.2.1.jar
  xwork-core-2.2.1.jar
- Struts2 的 Spring 插件 Jar 包：

  struts2-convention-plugin-2.2.1.jar
  struts2-spring-plugin-2.2.1.jar

- Spring 必需的 Jar 包：
  spring-aop-4.1.0.RELEASE.jar
  spring-aspects-4.1.0.RELEASE.jar
  spring-beans-4.1.0.RELEASE.jar
  spring-context-4.1.0.RELEASE.jar
  spring-context-support-4.1.0.RELEASE.jar
  spring-core-4.1.0.RELEASE.jar
  spring-expression-4.1.0.RELEASE.jar
  spring-jdbc-4.1.0.RELEASE.jar
  spring-tx-4.1.0.RELEASE.jar
  spring-web-4.1.0.RELEASE.jar
  spring-webmvc-4.1.0.RELEASE.jar
- Hibernate 必需的 Jar 包：
  antlr-2.7.7.jar
  dom4j-1.1.6.jar
  hibernate-commons-annotations-4.0.1.Final.jar
  hibernate-core-4.1.4.Final.jar
  hibernate-jpa-2.0-api-1.0.1.Final.jar
  javassist-3.15.0-GA.jar
  jobss-logging-3.1.0.GA.jar
  jobss-transaction-api_1.1_spec-1.0.0. Final.jar
- JSTL 必需的 Jar 包：
  javax.servlet.jsp.jstl.jar
  jstl-impl-1.2.2.jar
- 为了简单，只删除同名但版本号较低的 Jar 包。清理完成后，剩余 64 个 Jar 包（其中有 9 个是手工放进去的），把 lib 下的所有 Jar 包复制到 MyEclipse 工具 newsssh 项目的 lib 文件夹中。
- 右击项目 newsssh，单击 Properties 中的 Java Build Path，在右边窗口中切换到 Libraries 选项卡，如图 4-21 所示，删除（Remove）Hibernate、JSTL、Spring 和 Struts 的 Libraries，如图 4-22 所示。

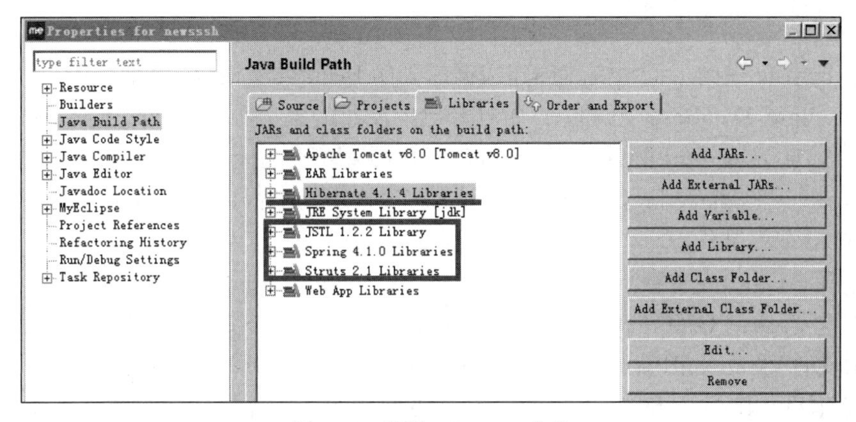

图 4-21　删除 Libraries 之前

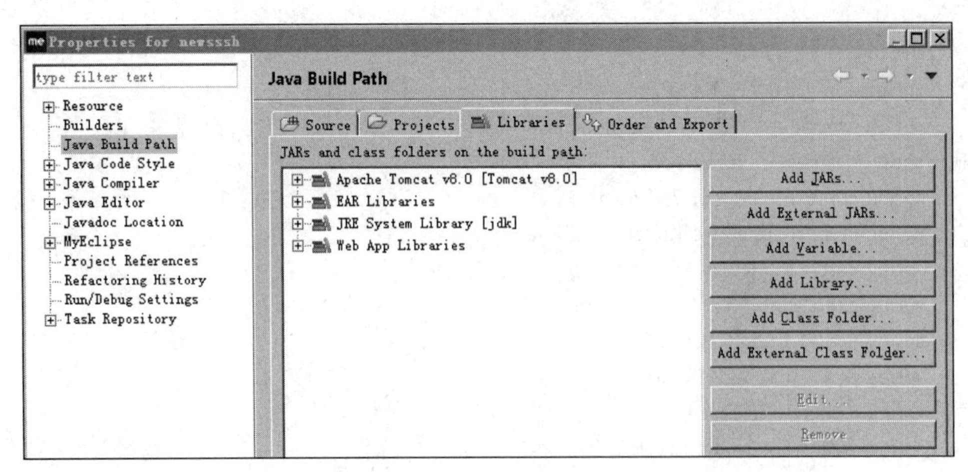

图 4-22 删除 Libraries 之后

- 重新运行当前项目，查看是否运行正常、是否可以正常显示项目首页。

4.2.12 考核任务

正确实现 SSH 项目的搭建。检查点：

（1）可以在浏览器看到默认首页 index.jsp 的内容，无乱码。（40 分）

（2）MyEclipse 的 Console 中没有报错（如 Exception）信息。（30 分）

（3）项目中已添加 SSH 三个框架的 Jar 包。（30 分）

4.3 日志系统

在 config 中添加 log4j.properties，配置内容与 3.4 节中相同，以便在调试时看到更多的日志。

4.4 创建 Bean 类及对应的 hbm 映射文件

4.4.1 Hibernate 逆向工程

借助 Hibernate 逆向工程，快速创建数据库表的 Bean 类和相应的 Hibernate 的描述文件。

在 MyEclipse DataBase Explorer 中打开已经创建的数据源连接 newsssh，右击需要创建 Bean 类的表，如 user 表、news 表和 cmenu 表，可以一次选择多个表，然后选择右键菜单项 Hibernate Reverse Engineering，如图 4-23 所示。设置逆向工程，如图 4-24 所示，生成需要的 Bean 类和 hbm 映射文件。

- 选择 Bean 类文件存放的 package 为 cn.lrw.newsssh.bean。
- 勾选 Create POJO<>DB Table mapping information、需要生成数据库表的 Hibernate 映射文件*.hbm.xml、自动更新 Hibernate 的配置（添加映射文件路径）。
- 勾选 Java Data Object，不勾选 Create abstract class。需要生成与数据库表对应的 Bean 类，不需要生成抽象 abstract 类。

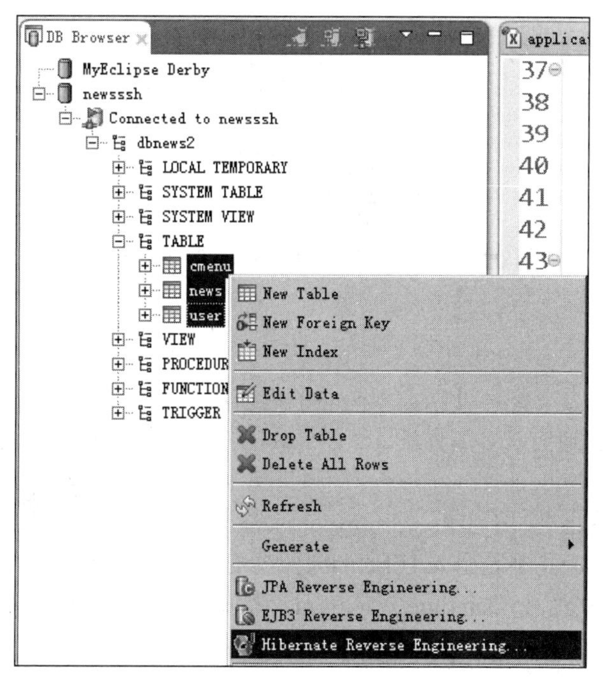

图 4-23　选择 Hibernate 逆向工程

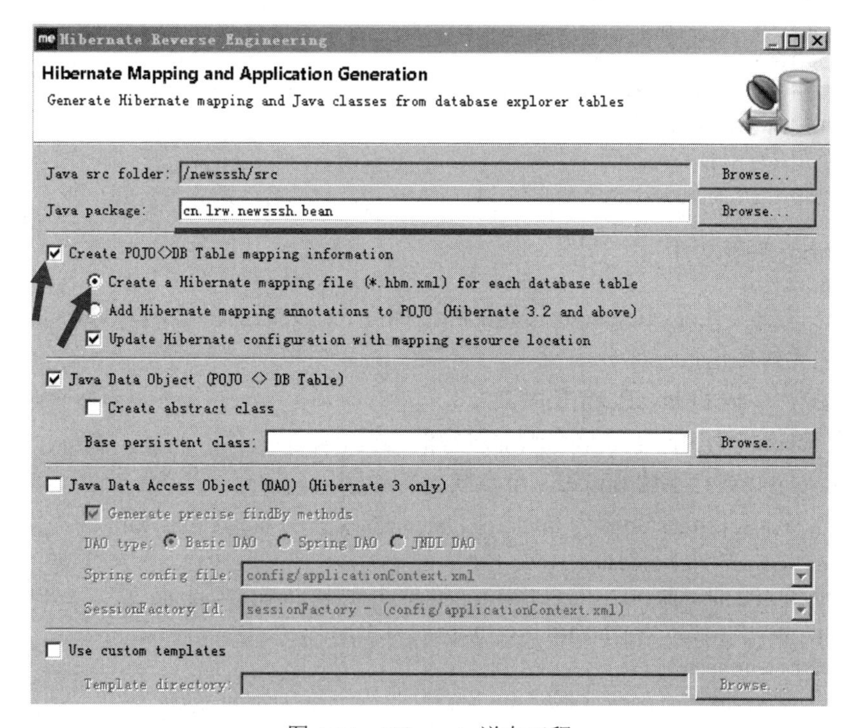

图 4-24　Hibernate 逆向工程

设置完成后，单击 Finish 按钮，然后在项目浏览器或 package 浏览器中，看到 cn.lrw.newsssh.bean 里面自动创建了 Bean 类和对应的 Hibernate 的映射文件，如图 4-25 所示。映射文件描述 Bean 类关联的数据库表、属性关联表的字段、数据类型等信息。

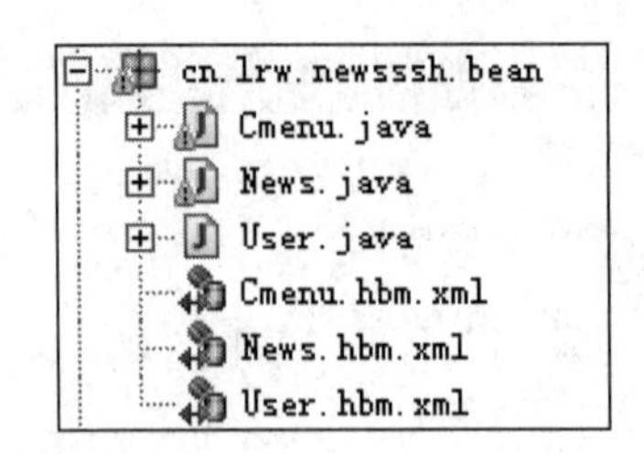

图 4-25 逆向工程生成的 Bean 类和 hbm 映射文件

4.4.2 Bean 类

打开生成的 Bean 类文件，可以看到文件中包括自动生成的 package 名、引入的类、Bean 类名、类中的属性、构造方法、属性的 getter/setter 方法。以 News.java 为例，还需要做少量的手工修改：

（1）在类名前面添加注解@Entity，标注这个类是一个实体 Bean，需要引入 javax.persistence.Entity，并不是引入 Hibernate 中的 Entity。

（2）默认与 MySQL 中 Datetime 对应的日期类型是 Timestamp，将其修改为 Date 类型，引入 java.util.Date。

```
@Entity
public class News implements java.io.Serializable {
    private Integer id;
    private String title;
    private String content;
    private Date tjdate;
    private String cruser;
    private Integer hitnum;
```

4.4.3 hbm 映射文件

Hibernate 映射文件是 Hibernate 与数据库进行持久化关联的桥梁，以 News.hbm.xml 为例，映射文件的主要内容包括三部分。

（1）数据库、表和 JavaBean 关联。

（2）在<class>节点下用 id 节点映射对应的主键。

（3）在<class>节点下用 property 节点映射普通字段。

建议修改 java.sql.timestamp 为 java.util.Date。修改后的 News.hbm.xml 的完整内容如下：

```
<?xml version="1.0" encoding="UTF-8"?>
<!DOCTYPE hibernate-mapping PUBLIC "-//Hibernate/Hibernate Mapping DTD 3.0//EN"
"http://www.hibernate.org/dtd/hibernate-mapping-3.0.dtd">
<hibernate-mapping>
    <class name="cn.lrw.newsssh.bean.News" table="news" catalog="dbnews2">
        <id name="id" type="java.lang.Integer">
            <column name="id" />
            <generator class="identity" />
        </id>
        <property name="title" type="java.lang.String">
```

```
            <column name="title" length="100" not-null="true" />
        </property>
        <property name="content" type="java.lang.String">
            <column name="content" length="65535" not-null="true" />
        </property>
        <property name="tjdate" type="java.util.Date">
            <column name="tjdate" length="19" not-null="true" />
        </property>
        <property name="cruser" type="java.lang.String">
            <column name="cruser" length="50" not-null="true" />
        </property>
        <property name="hitnum" type="java.lang.Integer">
            <column name="hitnum" not-null="true" />
        </property>
    </class>
</hibernate-mapping>
```

4.4.4 Hibernate 配置

逆向工程完成后，applicationContext.xml 文件的 sessionFactory Bean 中，已自动添加了*.hbm.xml 路径的配置，如下所示：

```
<bean id="sessionFactory"
    class="org.springframework.orm.hibernate4.LocalSessionFactoryBean">
    <property name="dataSource">
        <ref bean="dataSource" />
    </property>
    <property name="hibernateProperties">
        <props>
            <prop key="hibernate.dialect">
                org.hibernate.dialect.MySQLDialect
            </prop>
        </props>
    </property>
    <property name="mappingResources">
        <list>
            <value>cn/lrw/newsssh/bean/News.hbm.xml</value>
            <value>cn/lrw/newsssh/bean/Cmenu.hbm.xml</value>
            <value>cn/lrw/newsssh/bean/User.hbm.xml</value></list>
    </property></bean>
```

4.5　封装 Tree 型数据

前端树形菜单将使用 EasyUI 的 Tree 控件，所以要在后端定义相应的 Bean 类 EasyUITree，由于没有对应的数据库表，不能用逆向工程生成，需要手工添加。当写完属性之后，在代码编辑窗口中，单击右键，选择 Source 中的 Generate Getters and Setters 命令，如图 4-26 所示，在弹出的窗口中，选择属性，单击 OK 按钮，将自动生成所选属性的 getter/setter 方法。部分代码如下：

```
public class EasyUITree {
    private String id;
    private String text;
    private Boolean checked = false;
    private Map<String,Object> attributes;
    private String state = "closed";
    public List<EasyUITree> children;
   //省略了部分 getter/setter 方法
    public String getId() {
        return id;
    }
    public void setId(String id) {
        this.id = id;
    }
}
```

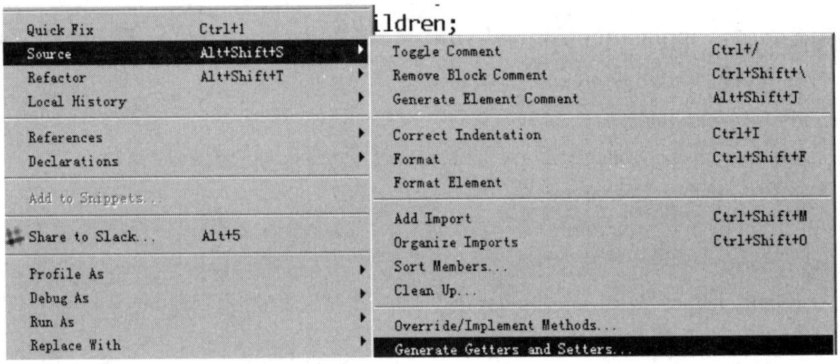

图 4-26　getter/setter 方法

4.6　封装 DAO

完成持久层的 Bean 类和 Hibernate 映射文件后，需要封装 DAO 来实现对持久层的访问。在 cn.lrw.newsssh.dao 包中创建 BaseDao 类，根据项目需要，封装一些访问持久层的方法。

其中用到几个注解：

- @Repository，将数据访问层（DAO 层）的类标识为 Spring Bean。
- @SuppressWarnings("all")，让编译器对被注解的代码元素内部的某些警告保持静默。
- @Autowired，Spring 框架中完成对成员变量、方法或构造函数自动装配的工作，不需要 getter()和 setter()方法，Spring 会自动注入。

封装 DAO 时用到泛型来实现方法的复用，减少重复代码的编写，提高开发效率，另一方面，无论访问持久层的哪一个实体类，都可以使用同一套方法。

DAO 中用到一个非常重要的对象 session。方法 getCurrentSession 获取当前上下文一个 session 对象。当第一次使用此方法时，会自动产生一个 session 对象；连续使用多次时，得到的 session 都是同一个对象。调用方法 getCurrentSession 之前，必须在 Spring 中已经正确配置 sessionFactory，才能让 Spring 管理 Hibernate 的事务，有了事务管理，才能正常使用对象 session。

```
@Repository
@SuppressWarnings("all")
public class BaseDao<T>   {
    private SessionFactory sessionFactory;
    public SessionFactory getSessionFactory() {
        return sessionFactory;
    }
    @Autowired
    public void setSessionFactory(SessionFactory sessionFactory) {
        this.sessionFactory = sessionFactory;
    }
    private Session getCurrentSession() {
        return sessionFactory.getCurrentSession();
}
```

下面是封装了增删改查四种操作的一些方法，代码简单易用。

4.6.1　增

保存一个对象，也就是保存一条记录到数据库表中。

```
public Serializable save(T o) {
    return this.getCurrentSession().save(o);
}
```

4.6.2　删

删除一个对象，也就是从数据库表中删除一条记录。

```
public void delete(T o) {
    this.getCurrentSession().delete(o);
}
```

4.6.3　改

更新一个对象，也就是更新数据库表中的一条记录。

```
public void update(T o) {
    this.getCurrentSession().update(o);
}
```

4.6.4　查

对数据库的查询操作使用率特别高，所以查询的需求也相对比较多。可以封装多个不同的方法，实现查询单个对象、多个对象、对象的数量，或者其他特殊的查询。使用 Java 重载，同一个方法名有不同的参数，实现不同需求（提供了不同的查询参数）的同一查询目标（查询结果为单个对象，或对象集合）。需要特别提醒的是，方法可能使用了 HQL（Hibernate Query Language），这是因为当前项目使用了 Hibernate 框架。HQL 与 SQL 有一定的区别，在业务逻辑类中写 HQL 时尤其要注意其语法结构。

- HQL 面向对象查询，SQL 面向数据库查询。
- HQL 的语法结构：from +类名+类对象 +where+类对象属性的条件。
- SQL 的语法结构：from +数据库表名 + where+表字段条件。
- HQL 不需要 insert 语句，只需构造新增对象后调用 save()方法。
- HQL 不需要 update 语句，只需得到修改对象后调用 update()方法。
- HQL 不需要 delete 语句，只需得到删除对象后调用 delete()方法。

1. 查询单个对象

```
public T get(Class<T> c, Serializable id) {
    return (T) this.getCurrentSession().get(c, id);
}
public T get(String hql, Object[] param) {
    List<T> l = this.find(hql, param);
    if (l != null && l.size() > 0) {
        return l.get(0);
    } else {
        return null;
    }
}
public T get(String hql, List<Object> param) {
    List<T> l = this.find(hql, param);
    if (l != null && l.size() > 0) {
        return l.get(0);
    } else {
        return null;
    }
}
```

2. 查询多个对象

查询多个对象，大部分情况得到的是一个对象集合。

```
public List<T> find(String hql) {
    return this.getCurrentSession().createQuery(hql).list();
}
public List<T> find(String hql, Object[] param) {
    Query q = this.getCurrentSession().createQuery(hql);
    if (param != null && param.length > 0) {
        for (int i = 0; i < param.length; i++) {
            q.setParameter(i, param[i]);
        }
    }
    return q.list();
}
public List<T> find(String hql, List<Object> param) {
    Query q = this.getCurrentSession().createQuery(hql);
    if (param != null && param.size() > 0) {
```

```
            for (int i = 0; i < param.size(); i++) {
                q.setParameter(i, param.get(i));
            }
        }
            return q.list();
    }
```

很多前端控件需要带分页功能显示列表，如分页显示新闻列表，所以查询时带分页参数：page 表示查询第几页，rows 表示每页显示多少条记录。方法内部对参数 page 和 rows 做了默认处理，page 异常时，默认为 1；rows 异常时，默认为 10。

```
    public List<T> find(String hql, Object[] param, Integer page, Integer rows) {
        if (page == null || page < 1) {
            page = 1;
        }
        if (rows == null || rows < 1) {
            rows = 10;
        }
        Query q = this.getCurrentSession().createQuery(hql);
        if (param != null && param.length > 0) {
            for (int i = 0; i < param.length; i++) {
                q.setParameter(i, param[i]);
            }
        }
        return q.setFirstResult((page - 1) * rows).setMaxResults(rows).list();
    }
    public List<T> find(String hql, List<Object> param, Integer page, Integer rows) {
        if (page == null || page < 1) {
            page = 1;
        }
        if (rows == null || rows < 1) {
            rows = 10;
        }
        Query q = this.getCurrentSession().createQuery(hql);
        if (param != null && param.size() > 0) {
            for (int i = 0; i < param.size(); i++) {
                q.setParameter(i, param.get(i));
            }
        }
        return q.setFirstResult((page - 1) * rows).setMaxResults(rows).list();
    }
```

3. 查询对象数量

参数 hql 的写法默认为：select count(*) from 类名

```
        public Long count(String hql) {
            try{
            return (Long) this.getCurrentSession().createQuery(hql).uniqueResult();
            }catch(Exception e){
```

```
            e.printStackTrace();
                return 0L;
            }
        }
        public Long count(String hql, Object[] param) {
            Query q = this.getCurrentSession().createQuery(hql);
            if (param != null && param.length > 0) {
                for (int i = 0; i < param.length; i++) {
                    q.setParameter(i, param[i]);
                }
            }
            return (Long) q.uniqueResult();
        }
        public Long count(String hql, List<Object> param) {
            Query q = this.getCurrentSession().createQuery(hql);
            if (param != null && param.size() > 0) {
                for (int i = 0; i < param.size(); i++) {
                    q.setParameter(i, param.get(i));
                }
            }
            return (Long) q.uniqueResult();
    }
```

4. 特殊 HQL 语句的执行

执行特殊的HQL语句，返回受影响的对象数量。

```
public Integer executeHql(String hql) {
    return this.getCurrentSession().createQuery(hql).executeUpdate();
}
public Integer executeHql(String hql, Object[] param) {
    Query q = this.getCurrentSession().createQuery(hql);
    if (param != null && param.length > 0) {
        for (int i = 0; i < param.length; i++) {
            q.setParameter(i, param[i]);
        }
    }
    return q.executeUpdate();
}
public Integer executeHql(String hql, List<Object> param) {
    Query q = this.getCurrentSession().createQuery(hql);
    if (param != null && param.size() > 0) {
        for (int i = 0; i < param.size(); i++) {
            q.setParameter(i, param.get(i));
        }
    }
    return q.executeUpdate();
}
```

4.7　公共方法类

在业务逻辑类和控制器类中可能多次用到一些相同的方法，开发人员通常会把这样的方法封装成工具类 BaseUtil，存放于 cn.lrw.newsssh.utils 包中，作为公共方法类。当前的公共方法介绍如下。

4.7.1　字符串加密

lrwCode 方法，把字符串用 shiro 提供的 Sha256Hash 加密。

```
public static String lrwCode(String password,String salt){
    if(salt==""){
        salt="abcdefghijklmnopqrstuvwx";
    }
    return new Sha256Hash(password, salt, 1024).toBase64();
}
```

4.7.2　字符串输出

把控制器处理的字符串结果输出到前端。

```
public static void outPrint(HttpServletResponse response, String data) {
    response.setContentType("application/json;charset=UTF-8");
    response.setCharacterEncoding("UTF-8");
    try {
        PrintWriter out = response.getWriter();
        out.print(data);
        out.flush();
        out.close();
    } catch (Exception e) {
        System.out.println(e.getMessage());
    }
}
```

4.7.3　字符串判断

判断字符串是否为空，或者是否除空格没有其他字符。

```
public static boolean isNull(String str) {
    if (str != null && !str.trim().equals("")) {
        return false;
    } else {
        return true;
    }
}
```

4.7.4　对象与 JSON 串相互转换

Gson 是 Google 提供的用来在 Java 对象和 JSON 数据之间进行映射的 Java 类库。可以将一个 JSON 字符转成一个 Java 对象，或者将一个 Java 对象转化为 JSON 字符串，特点是快速高效、代码量少、简洁、面向对象、数据传递和解析方便。

1. 定义 Gson 对象

首先定义 Gson 的单例对象。

```
private static Gson gson=new Gson();
```

2. 对象转 JSON 串

```
public static String toJson(Object src) {
    if (src == null) {
        return gson.toJson(JsonNull.INSTANCE);
    }
    return gson.toJson(src);
}
```

3. JSON 串转对象

```
public static <T> Object fromJson(String json, Class<T> classOfT) {
    return gson.fromJson(json, (Type) classOfT);
}
public static Object fromJson(String json, Type typeOfT) {
    return gson.fromJson(json, typeOfT);
}
```

4.8 自定义 Filter

项目中将使用百度在线编辑器 UEditor，上传文件的请求会被 Struts2 的过滤器默认拦截，所以在 cn.lrw.newsssh.utils 中自定义一个 Filter，假定命名为 MyStrutsFilter，替代 Struts2 的过滤器 StrutsPrepareAndExecuteFilter，放开对 UEditor 请求上传的拦截。MyStrutsFilter 继承了 StrutsPrepareAndExecuteFilter，重载了 doFilter 方法，方法中判断前端请求的 URL 中出现 controller.jsp 时，直接跳过不拦截，执行下一个过滤器 Filter。

替代的方法是修改 web.xml 默认的 Struts2 过滤器：

修改前：org.apache.struts2.dispatcher.ng.filter.StrutsPrepareAndExecuteFilter

修改后：cn.lrw.newsssh.utils.MyStrutsFilter

MyStrutsFilter 的代码如下：

```
public class MyStrutsFilter extends StrutsPrepareAndExecuteFilter {
    @Override
    public void doFilter(ServletRequest req, ServletResponse res,FilterChain chain) throws
IOException, ServletException {
        HttpServletRequest request = (HttpServletRequest) req;
        String url = request.getRequestURI();
        req.setCharacterEncoding("UTF-8");
        res.setCharacterEncoding("UTF-8");
        if (url.contains("controller.jsp")) {
            System.out.println("myfilter:"+url);
            chain.doFilter(req, res);
        }else{
```

```
            super.doFilter(req, res, chain);
        }
    }
}
```

4.9　创建业务逻辑类

编写业务逻辑类，调用 DAO 操作，其中用到如下几个注解：

- @Service，标注业务层组件。
- @Resource，把 DAO 注入到 service 中，不再需要 new 一个对象。

4.9.1　UserSvc 类

在 cn.lrw.examssh.service 包中创建 UserSvc 类，实现添加新用户、查询用户等业务逻辑。

```
@Service
public class UserSvc {
    @Resource
    private BaseDao<User> dao;
    public void addU(User user) {
        dao.save(user);
    }
    public User findU(String uid, String pwd) {
        return dao.get(" from User u where u.uid = ? and u.pwd = ? ", new Object[]
            { uid, BaseUtil.lrwCode(pwd, "") });
    }
    public Long getCount() {
        return dao.count("select count(*) from User");
    }
}
```

4.9.2　NewsSvc 类

在 cn.lrw.examssh.service 包中创建 NewsSvc 类，实现新闻信息增删改查的业务逻辑。

```
@Service
public class NewsSvc {
    @Resource
    private BaseDao<News> dao;
    //===========增===============
    public void addNews(News news) throws Exception{
        dao.save(news);
    }
    //===========删===============
    public void deleteNews(int id,Class<News> news) throws Exception{
        News u =(News) dao.get(news, id);
        dao.delete(u);
    }
```

```
    //==========改==============
    public void updateNews(News news) throws Exception{
        dao.update(news);
    }
    //==========查==============
    //按新闻标题分页查询
    public List<News> listDgNews(String title,int page,int rows){
        if(title == null || "".equals(title)) return dao.find("from News news order by news.tjdate desc",
new Object[0], page, rows);
        else return dao.find("from News news WHERE news.title like ? order by news.tjdate desc",
new Object[]{'%' +title+'%'}, page, rows);
    }
    //按 id 查询/阅读新闻
    public News getNews(Class<News> clazz, int id){
        News news=dao.get(clazz, id);
        news.setHitnum(news.getHitnum()+1);//单击量增加
        dao.update(news);
        return   news;
    }
    //新闻记录数量
    public int getNewsCount(){
        try{
            Long a=dao.count("select count(*) from News");
        return   Integer.parseInt(a.toString());
        }catch(Exception e){
            e.printStackTrace();
            return 0;
        }
    }
}
```

4.9.3 MenuSvc 类

在 cn.lrw.examssh.service 包中创建 MenuSvc 类，实现 Tree 形菜单数据的获取。

```
@Service
public class MenuSvc {
    @Resource
    private BaseDao<Cmenu> dao;
    public List<Cmenu> listMenu(int pid){
        return dao.find("from Cmenu cmenu where cmenu.pid=?", new Object[]{pid});
    }
}
```

4.10 创建控制器类

编写控制器类处理前端页面的请求，实现与业务逻辑的交互，进行一些必要的数据验证、转换，用到如下的几个注解：

@Controller，自动根据 Bean 的类名实例化一个首写字母为小写的 Bean，如 userAct。

@Resource，把 service 注入到 action 中，不需要新建（new）一个对象。

4.10.1　UserAct 类

在 cn.lrw.examssh.action 包中创建 UserAct 类，实现用户登录、退出和跳转到后台的请求。

在 Struts2 中，底层的 session 都被封装成了 Map 类型的 SessionMap，得到 SessionMap 之后，就可以对 session 进行读写，不必再去使用底层的 session。

```
Map<String, Object> session = ActionContext.getContext().getSession();
```

通常意义上的 session 是指 HttpSession。如果想得到原始的 HttpSession，可以先得到 HttpServletRequest 对象，再通过 request.getSession()来取得原始的 HttpSession 对象。

如果需要接收前端传过来的参数值，可以在当前类里面定义与参数名同名的属性，如 uid、pwd，然后给它们添加相应的 getter/setter 方法。如果参数是对象，处理方法类似。

```
@Controller
public class UserAct extends ActionSupport {
    @Resource
    private UserSvc userSvc;
    private String uid,pwd;
    private String jsonResult;
    private HashMap<String,Object> jsonobj=new HashMap<String,Object>();
    public String doLogin(){
        try {
            Long n=userSvc.getCount();
            if(n==0){
                User user=new User();
                user.setUid("2953");
                user.setXm("lrw");
                user.setPwd(BaseUtil.lrwCode("123", ""));
                user.setRole("1");
                userSvc.addU(user);
            }
            User user0 = userSvc.findU(uid, pwd);
            jsonobj.clear();
            if(user0 != null){
                jsonobj.put("ok", true);
                jsonobj.put("msg", "goIndex");
                Map<String, Object> session = ActionContext.getContext().getSession();
                session.put("me", user0);
            }else
                {
                jsonobj.put("ok", false);
                jsonobj.put("msg", "用户不存在");
                }
        } catch (Exception e) {
```

```
                jsonobj.put("ok", false);
                jsonobj.put("msg", "系统错误 2");
            }
            jsonResult = BaseUtil.toJson(jsonobj);
            HttpServletResponse response = ServletActionContext.getResponse();
            BaseUtil.outPrint(response, jsonResult);
            return null;
        }
        public String doLogout(){
            Map<String, Object> session = ActionContext.getSession();
            session.put("me", null);
            return "logout";
        }
        //请求跳转到新闻管理页
        public String goIndex(){
            return "goadmin";
        }
        //属性的 get/set 方法
        public String getUid() {
            return uid;
        }
        public void setUid(String uid) {
            this.uid = uid;
        }
        public String getPwd() {
            return pwd;
        }
        public void setPwd(String pwd) {
            this.pwd = pwd;
        }
    }
```

4.10.2 NewsAct 类

在 cn.lrw.examssh.action 包中创建 NewsAct 类，实现对新闻进行增删改查的请求。@SuppressWarnings("serial")是关闭序列化的警告。

（1）前端页面传过来的参数有：

- 一条新闻 news。
- 一条新闻的 id。
- 分页参数 page 和 rows。
- 查询新闻标题的关键词 s_name。

（2）jsonobj 是自定义的一个 Map 对象，用于封装需要返回的数据信息，每次使用前最好做清空处理：jsonobj.clear()。

（3）jsonResult 是将封装的结果转换到 JSON 字符串，用于返回到前端。

```
@SuppressWarnings("serial")
@Controller("news")
public class NewsAct extends ActionSupport {
	@Resource
	private NewsSvc newsSvc;
	private News news;
	private int page,rows,id;
	private String s_name;
	private String jsonResult;
	private HashMap<String,Object> jsonobj=new HashMap<String,Object>();
//===========增===============
	//请求跳转到添加新闻页面
	public String goAdd(){
		return "goadd";
	}
	//保存添加的新闻
	public String saveAdd(){
		jsonobj.clear();
		try {
			news.setTjdate(new Date());//提交日期由后端生成
			news.setHitnum(0);
			newsSvc.addNews(news);
			jsonobj.put("ok", true);
			jsonobj.put("msg", "goadmin");
		} catch (Exception e) {
			e.printStackTrace();
			jsonobj.put("ok", false);
			jsonobj.put("msg", "系统错误 2");
		}
		jsonResult = BaseUtil.toJson(jsonobj);
		HttpServletResponse response = ServletActionContext.getResponse();
		BaseUtil.outPrint(response, jsonResult);
		return null;
	}
//===========删===============
	//删除一条新闻
	public String doDel1(){
		jsonobj.clear();
		boolean delflag=false;
		try{
			newsSvc.deleteNews(id, News.class);
			delflag=true;
		}catch(Exception e){
			e.printStackTrace();
			delflag=false;
		}
```

```
            jsonobj.put("delflag", delflag);
            HttpServletResponse response = ServletActionContext.getResponse();
            BaseUtil.outPrint(response, BaseUtil.toJson(jsonobj));
            return null;
        }
//===========改================
        //请求跳转到修改新闻页
        public String goEdit(){
            news=newsSvc.getNews(News.class, id);
            return "goedit";
        }
        //保存修改后的新闻
        public String saveEdit(){
            jsonobj.clear();
            try {
                News news0=newsSvc.getNews(News.class, news.getId());
                news0.setContent(news.getContent());
                news0.setCruser(news.getCruser());
                news0.setTitle(news.getTitle());
                newsSvc.updateNews(news0);
                jsonobj.put("ok", true);
                jsonobj.put("msg", "goadmin");
            } catch (Exception e) {
                e.printStackTrace();
                jsonobj.put("ok", false);
                jsonobj.put("msg", "系统错误 2");
            }
            jsonResult = BaseUtil.toJson(jsonobj);
            HttpServletResponse response = ServletActionContext.getResponse();
            BaseUtil.outPrint(response, jsonResult);
            return null;
        }
//===========查================
        //请求跳转到后台新闻列表页
        public String goList(){
            return "golist";
        }
        //统计新闻总数量
        public String getCount(){
            int c=0;
            try{
                c=newsSvc.getNewsCount();
            }catch(Exception e){
                e.printStackTrace();
                c=0;
            }
```

```
        jsonobj.clear();
        jsonobj.put("newscount", c);
        jsonResult = BaseUtil.toJson(jsonobj);
        HttpServletResponse response = ServletActionContext.getResponse();
        BaseUtil.outPrint(response, jsonResult);
        return null;
    }
    //阅读一条新闻
    public String getaNews(){
        news=newsSvc.getNews(News.class, id);
        return "goread";
    }
    //分页查询新闻
    public String listNews(){
        List<News> newslist=newsSvc.listDgNews(s_name,page,rows);
        jsonobj.clear();
        jsonobj.put("total", newslist.size());
        jsonobj.put("rows", newslist);
        jsonResult = BaseUtil.toJson(jsonobj);
        HttpServletResponse response = ServletActionContext.getResponse();
        BaseUtil.outPrint(response, jsonResult);
        return null;
    }
//===========属性的 get/set 方法===============
    public News getNews() {
        return news;
    }
    public void setNews(News news) {
        this.news = news;
    }
    public String getS_name() {
        return s_name;
    }
    public void setS_name(String s_name) {
        this.s_name = s_name;
    }
    public int getPage() {
        return page;
    }
    public void setPage(int page) {
        this.page = page;
    }
    public int getRows() {
        return rows;
    }
    public void setRows(int rows) {
```

```
            this.rows = rows;
        }
        public int getId() {
            return id;
        }
        public void setId(int id) {
            this.id = id;
        }
    }
```

4.10.3 MenuAct 类

在 cn.lrw.examssh.action 包中创建 MenuAct 类，实现对菜单数据的封装。

```
    @SuppressWarnings("serial")
    @Controller
    public class MenuAct extends ActionSupport {
        @Resource
        private MenuSvc menuSvc;
        private String jsonResult;
        public String menutree() {
            Map<String, Object> session=ActionContext.getContext().getSession();
            User user = (User) session.get("me");
            String role=user.getRole();
            //=========一级菜单=================
            List<Cmenu> menulist=menuSvc.listMenu(0);
            List<EasyUITree> eList = new ArrayList<EasyUITree>();
            if(menulist.size() != 0){
                for (int i = 0; i < menulist.size(); i++) {
                    Cmenu t = menulist.get(i);
                    if(!t.getPermission().contains(role))continue;
                    EasyUITree e = new EasyUITree();
                    e.setId(t.getId()+"");
                    e.setText(t.getName());
                    List<EasyUITree> eList2 = new ArrayList<EasyUITree>();
                //==========二级菜单==============
                    List<Cmenu> menu2 = menuSvc.listMenu(t.getId());
                    for (int j = 0; j < menu2.size(); j++) {
                        Cmenu t2 = menu2.get(j);
                        if(!t2.getPermission().contains(role))continue;
                        Map<String,Object> attributes = new HashMap<String, Object>();
                        attributes.put("url", t2.getUrl());
                        attributes.put("role", t2.getPermission());
                        EasyUITree e1 = new EasyUITree();
                        e1.setAttributes(attributes);
                        e1.setId(t2.getId()+"");
                        e1.setText(t2.getName());
                        e1.setState("open");
```

```
                    eList2.add(e1);
                }
                e.setChildren(eList2);
                e.setState("closed");
                eList.add(e);
            }
        }
        jsonResult = BaseUtil.toJson(eList);
        HttpServletResponse response = ServletActionContext.getResponse();
        BaseUtil.outPrint(response, jsonResult);
        return null;
    }
}
```

4.11　配置 Spring

在 applicationContext.xml 文件中已经自动配置了数据源 dataSource、会话工厂 sessionFactory、事务管理 transactionManager。由于在类中使用了注解，在这个配置文件中不再需要定义 DAO，Service 和 Action 的 Bean。

还需要配置事务通知属性 txAdvice；定义事务传播属性 tx:attributes；配置事务切面 AOP 只对业务逻辑层实施事务；配置包的自动扫描 base-package 为需要扫描的包 cn.lrw.newsssh，包含所有子包 Action、Service、DAO、Bean 和 Utils，启动了包扫描，将 cn.lrw.newsssh 这个包下以及子包下的所有类扫描一遍，将标记有@Controller、@Service、@repository、@Component 等注解的类注入到 IOC 容器中，作为 Spring 的 Bean 来管理。

在 applicationContext.xml 文件中添加的配置信息如下：

```
    <!-- 配置事务通知属性 -->
<tx:advice id="txAdvice" transaction-manager="transactionManager">
    <!-- 定义事务传播属性 -->
    <tx:attributes>
    <!-- 只有 save、delete、update 等开头的方法才能执行增删改查操作 -->
        <tx:method name="insert*" propagation="REQUIRED" />
        <tx:method name="update*" propagation="REQUIRED" />
        <tx:method name="edit*" propagation="REQUIRED" />
        <tx:method name="save*" propagation="REQUIRED" />
        <tx:method name="add*" propagation="REQUIRED" />
        <tx:method name="new*" propagation="REQUIRED" />
        <tx:method name="set*" propagation="REQUIRED" />
        <tx:method name="remove*" propagation="REQUIRED" />
        <tx:method name="delete*" propagation="REQUIRED" />
        <tx:method name="change*" propagation="REQUIRED" />
        <tx:method name="get*" propagation="REQUIRED"    />
        <tx:method name="find*" propagation="REQUIRED" read-only="true" />
        <tx:method name="load*" propagation="REQUIRED" read-only="true" />
        <tx:method name="*" propagation="REQUIRED" read-only="true" />
```

```
        </tx:attributes>
    </tx:advice>
        <!-- 配置事务切面 AOP ,只对业务逻辑层实施事务-->
    <aop:config>
        <aop:pointcut id="txPointcut" expression="execution(* cn.lrw.newsssh.service..*.*(..))" />
        <!-- Advisor 定义，切入点和通知分别为 txPointcut、txAdvice -->
        <aop:advisor advice-ref="txAdvice" pointcut-ref="txPointcut" />
    </aop:config>
    <!-- 自动扫描包-->
    <context:component-scan base-package="cn.lrw.newsssh" />
```

由于添加了 aop 和 context 标签，需要在 beans 节点中添加标签所需的命名空间、描述该命名空间的模式文档所在的具体位置。如图 4-27 所示，在 package 浏览器中，右键单击 applicationContext.xml 文件，选择 Open With 中的 MyEclipse Spring Config Editor 命令，打开 Spring 配置文件专用编辑器，切换到 Namespaces 栏，如图 4-28 所示，勾选需要添加的标签所对应的 XSD 命名空间。

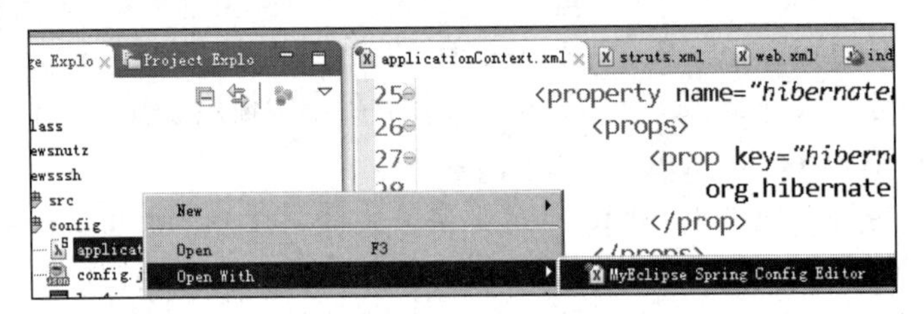

图 4-27 打开 Spring 配置文件

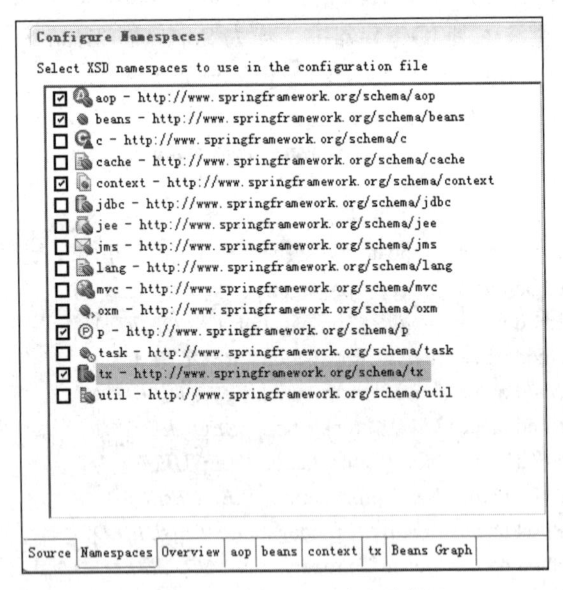

图 4-28 选择 XSD 命名空间

添加 XSD 命名空间后，applicationContext.xml 文件的根节点 beans 变成下述代码段的内容，其中 xmlns 是指 XML 命名空间（Namespace）。由于 XML 文件的标签名大都是自定义的，为防止不同人定义的标签命名冲突，所以需要加上一个 Namespace 来区分不同的 XML 文件，类似于 Java 中的 package；xmlns:xsi（xml schema instance）是指 XML 文件遵守的 XML 规范；xsi:schemaLocation 是指本文档里的 XML 元素所遵守的规范和引用的模式文档，它的值是成对出现的，第一个值表示命名空间，第二个值则表示描述该命名空间的模式文档的具体位置，两个值之间以空格分隔，解析器可以在需要的情况下使用这个文档对 XML 实例文档进行校验。

```
<beans xmlns="http://www.springframework.org/schema/beans"
    xmlns:xsi="http://www.w3.org/2001/XMLSchema-instance"
    xmlns:p="http://www.springframework.org/schema/p"
    xmlns:tx="http://www.springframework.org/schema/tx"
    xmlns:context="http://www.springframework.org/schema/context"
    xmlns:aop="http://www.springframework.org/schema/aop"
    xsi:schemaLocation="http://www.springframework.org/schema/beans
http://www.springframework.org/schema/beans/spring-beans-4.1.xsd
        http://www.springframework.org/schema/context
http://www.springframework.org/schema/context/spring-context-2.5.xsd
        http://www.springframework.org/schema/aop
http://www.springframework.org/schema/aop/spring-aop-2.5.xsd
        http://www.springframework.org/schema/tx
http://www.springframework.org/schema/tx/spring-tx.xsd">
```

4.12　配置 Struts

struts.xml 是 SSH 框架项目的核心配置文件之一，主要包括以下元素：

- action 配置用户请求和相应 Action 之间的对应关系。
- result 配置可能用到的参数和返回结果。
- package 配置包。
- global-results 配置全局结果。
- interceptors 配置拦截器。
- include 导入其他配置文件。

4.12.1　配置 constant

constant 元素用于设置一些 Struts2 的常量，用于控制 Struts2 的一些特性和一些默认行为。比如：

- 指定 Web 应用的默认编码集为 UTF-8，该属性对于处理中文请求参数非常有用，相当于调用 HttpServletRequest 的 setCharacterEncoding 方法。
- 设置 Struts2 应用是否使用开发模式。如果项目在开发阶段，则设置该属性为 true，可以在应用出错时显示更多、更友好的出错提示；当进入项目发布阶段后，则设置该属性为 false。
- 指定上传文件的临时保存路径为当前项目根目录下的 tmp 文件夹。

把 Struts2 的 action 交给 Spring 去管理和注入属性。

```
<constant name="struts.i18n.encoding" value="UTF-8" />
<constant name="struts.devMode" value="false" />
```

```
<constant name="struts.multipart.saveDir" value="/tmp"/>
<constant name="com.opensymphony.xwork2.objectFactory" value="spring" />
```

4.12.2 配置 package

package 元素定义了一个包，在 Struts2 框架中，用包来组织 Action 和拦截器，每个包由零到多个拦截器及 Action 组成，便于模块化开发。

package 元素用来配置包。package 元素的可用属性如下：

- name，必选属性，用来唯一标识包，方便在其他包中被引用。
- extends，可选属性，让一个包继承另一个包，继承包可以从被继承包那里继承到拦截器、action 等，默认继承为“struts-default”。
- namespace，可选属性，可以指定包对应的命名空间，当指定后，该包所包含的 Action 处理的 URL 是：命名空间+action，默认值是“/”。

例如，设置包名 newsssh，由于控制器返回的数据已封装为 JSON 格式，所以设置继承 json-default。

```
<package name="newsssh" extends="json-default">
```

4.12.3 配置 global-results

global-results 元素配置包中的全局结果。当一个 action 处理用户请求后返回时，会首先在该 action 本身的局部结果中进行搜索，如果局部结果中没有对应的结果，则会查找全局结果。例如，当访问了不存在的页面时，就会自动转到/error/nologin.jsp 页面。

```
<global-results>
    <result name="error">/error/nologin.jsp</result>
    <result name="error500">/error/500.jsp</result>
</global-results>
```

4.12.4 配置 action 和 result

在配置 action 元素时，可用属性包括：

- name，必选属性，标识 action，指定该 action 所处理的请求的 URL，也就是 Action 类的映射名称。
- class，可选属性，指定 action 对象对应的实现类，如果控制器类中使用了注解 @Controller，该属性配置为 action 实例。默认情况，@Controller 注解会通知 Spring，被注解 Bean 类的实例名是一个首写字母为小写的 Bean 类名，如 userAct；如果 @Controller 注解有参数，则参数名为实例名，如@Controller("news")，news 是 NewsAct 的实例名；如果没有指定 class 属性值，则其默认值为类 com.opensymphony.xwork2.ActionSupport，它会使用默认的处理方法 execute()来处理请求。
- method，可选属性，指定请求 action 时调用的方法。如果指定 method 属性值，则该 action 调用 method 属性中指定的方法；否则调用默认的 execute()。如果 name 属性中使用通配符，method 属性中可以使用占位符来适配。

当调用 action 的 method 方法处理结束返回后，就是使用 result 元素来设置返回给浏览器的视图。配置 result 元素需指定 name 和 type 两个属性。

- name 属性，对应从 action 方法返回的值，默认值为 success。result 默认支持的 name 值有如下五种，当然可以自定义返回值。
  - SUCCESS：表示 Action 正确执行完成，返回相应的视图。
  - NONE：表示 Action 正确执行完成，但并不返回任何视图。
  - ERROR：表示 Action 执行失败，返回到错误处理视图。
  - INPUT：Action 的执行需要从前端界面获取参数，INPUT 就是代表这个参数输入的界面，一般在应用中，会对这些参数进行验证，如果验证没有通过，将自动返回到该视图。
  - LOGIN：Action 因为用户没有登录的原因没有正确执行，将返回该登录视图，要求用户进行登录验证。
- type 属性，指定结果类型，默认的类型是 dispatcher。result 默认支持的类型有很多种，以下是常用的几种。
  - dispatcher，请求转发，通常用于转向一个 JSP。
  - redirect，重定向到一个 URL。
  - chain，转发到另一个 action。
  - redirect- action，重定向到另一个 action。
  - plaintext，以原始文本显示，如显示文件源码。

如果返回 JSON 数据，则设置 result 的 type 为 JSON，还需要设置 root 元素，从根节点 jsonResult 开始遍历，才能获取真正的 JSON 数据。

```
<result type="json">
    <param name="root">jsonResult</param>
</result>
```

当前项目的 action 和 result 配置如下：

```
<!-- 用户登录退出后台 -->
    <action name="doLogin" class="userAct" method="doLogin">
        <result type="json">
            <param name="root">jsonResult</param>
        </result>
    </action>
    <action name="doLogout" class="userAct" method="doLogout">
        <result name="logout" type="redirect">/index.jsp</result>
    </action>
    <action name="goIndex" class="userAct" method="goIndex">
        <result name="goadmin">/WEB-INF/web/admin.jsp</result>
    </action>
<!-- 获取菜单 -->
    <action name="menu" class="menuAct" method="menutree">
        <result type="json">
            <param name="root">jsonResult</param>
        </result>
    </action>
```

```
<!-- 增加新闻 -->
    <action name="goAddNews" class="news" method="goAdd">
        <result name="goadd">/WEB-INF/web/newsadd.jsp</result>
    </action>
    <action name="saveAddNews" class="news" method="saveAdd">
        <result type="json">
            <param name="root">jsonResult</param>
        </result>
    </action>
<!-- 删除新闻 -->
    <action name="doDelNews" class="news" method="doDel1">
        <result type="json">
            <param name="root">jsonResult</param>
        </result>
    </action>
<!-- 修改新闻 -->
    <action name="goEditNews" class="news" method="goEdit">
        <result name="goedit">/WEB-INF/web/newsedit.jsp</result>
    </action>
    <action name="saveEditNews" class="news" method="saveEdit">
        <result type="json">
            <param name="root">jsonResult</param>
        </result>
    </action>
<!-- 查询新闻 -->
    <action name="goListNews" class="news" method="goList">
        <result name="golist">/WEB-INF/web/newslist.jsp</result>
    </action>
    <action name="getCountNews" class="news" method="getCount">
        <result type="json">
            <param name="root">jsonResult</param>
        </result>
    </action>
    <action name="getNews" class="news" method="getaNews">
        <result name="goread">/WEB-INF/web/newsread.jsp</result>
    </action>
    <action name="listNews" class="news" method="listNews">
        <result type="json">
            <param name="root">jsonResult</param>
        </result>
    </action>
```

4.13 前端页面

当前项目基于 SSH 框架，仍然以开发简易的新闻发布系统为例，旨在了解和实践 SSH 项目的搭建与实现，了解项目的工作流程。所以前端页面的主要内容与前一项目的页面内容基本一样，但由于当前项目没有使用 Beetl 模板引擎，页面是 JSP 格式，所以存在一些差别。

由于当前项目中的前端页面均使用 jsp 格式，如果通过向导创建 JSP 页面文件，则在文件开头默认定义了全局变量 basePath，表示当前项目的 URL，如 http://localhost:8080/newssh/，注意末尾的“/”。前端页面中引用 JS、CSS、图片等文件，或者向后端发送请求，均可以使用 basePath 变量，在 html 代码或者 JS 代码中的使用方式为：<%=basePath%>；如果 JS 是独立的文件，则需要在引用 JS 的 JSP 文件中，定义 JS 变量，如：<script>**var** base="<%=basePath%>";</script>，然后在 JS 文件中使用 base 变量。

```
<%@ page language="java" import="java.util.*" pageEncoding="UTF-8"%>
<%
String path = request.getContextPath();
String basePath = request.getScheme()+"://"+request.getServerName()+":"+request.getServerPort()+
path+"/";
%>
```

将第 3 章项目 include 里面的文件或文件夹全部复制到当前项目的 include 中。

在 WEB-INF\web 文件夹下新建 5 个 JSP 文件：admin.jsp、newsadd.jsp、newsedit.jsp、newslist.jsp、newsread.jsp。

在 error 文件夹中新建 nologin.jsp。

将第 3 章项目中同名前端文件的代码，分别替换当前相应 JSP 文件中的非 JSP 代码，然后稍作修改或调整，就可以完成基于 SSH 的项目前端页面的实现。

4.13.1　系统首页

index.jsp 是系统首页，分页显示新闻列表和登录窗口。首页位于 WebRoot 目录下，可以直接访问，所以可以使用相对路径引用 JS、CSS 文件和图片文件。详细的实现方法，请参照 3.5.1 节美化系统首页、3.5.2 节 Ajax 方法、3.5.3 节更友好的 alert、3.5.4 节标题图标、3.8.10 节新闻列表相关内容。

由下述的 index.jsp 的页面代码可见，文件的开头是 JSP 代码，定义了 basePath 变量。JS 中定义了变量 base，值就是 basePath 的值。

```
<%@ page language="java" import="java.util.*" pageEncoding="UTF-8"%>
<%
String path = request.getContextPath();
String basePath = request.getScheme()+"://"+request.getServerName()+":"+request.getServerPort()+ path+"/";
%>
<!DOCTYPE html PUBLIC "-//W3C//DTD XHTML 1.0 Transitional//EN" "http://www.w3.org/
TR/xhtml1/DTD/xhtml1-transitional.dtd">
<html xmlns="http://www.w3.org/1999/xhtml">
<head>
<meta http-equiv="Content-Type" content="text/html; charset=UTF-8" />
<title>简易新闻发布系统</title>
<link rel="shortcut icon" type="image/x-icon" href="./include/img/logo.png" />
<link rel="stylesheet" type="text/css" href="./include/css/main.css" />
<link rel="stylesheet" type="text/css" href="./include/easyui/themes/default/easyui.css" />
<link rel="stylesheet" type="text/css" href="./include/easyui/themes/icon.css" />
```

```
<script type="text/javascript" src="./include/js/jquery.min.js"></script>
<script type="text/javascript" src="./include/easyui/jquery.easyui.min.js"></script>
<script type="text/javascript" src="./include/easyui/locale/easyui-lang-zh_CN.js"></script>
<script>var base="<%=basePath%>";</script>
<script type="text/javascript" src="./include/js/login.js"></script>
</head>
<body>
<div style="float:right;padding-right:20px;">
<a id="a" href="#" style="margin-right:15px;" >登录</a>     <a id="b" href="#" >新闻</a>
</div>

<div class="login">
      <div id="llogin" class="box png">
            <div class="logo png"></div>
            <div class="input">
                  <div class="log" id="login_form">
                        <div class="name">
                              <label>用户名</label><input type="text" class="text" id="uid" placeholder="用
户名"   tabindex="1" />
                        </div>
                        <div class="pwd">
                              <label> 密     码 </label><input   type="password"   class="text"   id="pwd"
placeholder="密码"   tabindex="2" />
                              <input id="login_submit" type="button" class="submit" tabindex="3" value="登录" />
                              <div class="check"></div>
                        </div>
                        <div class="tip"></div>
                  </div>
            </div>
      </div>
<div id="lnews" class="l-wrap">
<div>
      <div>
            <div class="l-news">
                  <div class="nheader">
                        <table cellspacing="0" cellpadding="0"><tbody>
                              <tr><td><h3>通知新闻：</h3></td></tr>
                        </tbody></table>
                  </div>
                  <div class="nlist">
                        <table id="newstable" width="100%">
                          <tbody>
                              <tr id="trpp"><td colspan="3" align="left"> </td></tr>
                          </tbody></table>

                  </div>
```

```
                <div id="pp" style="background:#efefef;"></div>
            </div>
        </div>
    </div>
    </div>
        <div class="air-balloon ab-1 png"></div>
        <div class="air-balloon ab-2 png"></div>
        <div class="footer"></div>
    </div>
    </body>
    </html>
```

首页中引用了自定义的 login.js 文件。与第 3 章项目的 login.js 相比较，向后端请求的 URL 都发生了变化；分页请求新闻列表后，不再需要 res=JSON.parse(res)语句，代码如下。

```
var pageN=1,pageTotal=100;
$(function(){
    $('#login_form input').keydown(function (e) {
        if (e.keyCode == 13)
        {
            checkUserName();
        }
    });
    $("#login_submit").click(checkUserName);
    $.ajax({
        url:"./getCountNews",
        type:"post",
        success: function(res){
            pageTotal=parseInt(res);
            listNews(1,10);loadPager();
        },
        error:function(res){
            $.messager.alert("系统提示","系统错误","error");
        }
    });
    $("#llogin").hide();
    $("#a").click(function(){
        $("#llogin").show();
        $("#lnews").hide();
    });
    $("#b").click(function(){
        $("#llogin").hide();
        $("#lnews").show();
    });
});
function listNews(pageNumber,pageSize){
    $.ajax({
        url:"./listNews",
```

```
                data:{"page":pageNumber,"rows":pageSize},
                type:"post",
                success: function(res){
                    $(".inews").remove();
                    //res=JSON.parse(res);
                    if(res.total<=0){
                        var tr="<tr class='inews' height=\"25\"><td >";
                            tr+="<div class='t'>暂无相关新闻</div>";
                            tr+="</td><td width='1%' nowrap=''><span > </span></td></tr>";
                        $("#trpp").before(tr);
                    }
                    else {
                        pageN=pageNumber;
                        pageTotal=res.total;
                        var rows=res.rows;
                        for(var i=0;i<rows.length;){
                            var row=rows[i];
                            var tr="<tr class='inews' height=\"25\"><td >";
                                tr+="<div class='t'><a href='./getNews?id="+row.id+"' target='_blank'
                                title='"+row.title+"'>"+row.title+"</a></div>";
                                tr+="</td><td width='1%' nowrap=''><span >"+row.tjdate.substr(0,10)+
                                " </span></td></tr>";
                            $("#trpp").before(tr);
                            i++;
                            if(i%5==0){
                                tr="<tr class='inews' height='1'><td colspan='2' align='center'>";
                                  tr+="<hr style='border-style:dashed;border-color:#999999;width:99%;
                                  height:1px;border-width:1px 0px 0px 0px;visibility:inherit'></td></tr>";
                                $("#trpp").before(tr);
                            }
                        }
                    }
                },
                error:function(res){
                    $.messager.alert("系统提示","系统错误","error");
                }
            })
        }
        function loadPager(){
            $('#pp').pagination({
                total:pageTotal,
                pageSize:10,
                pageNumber:pageN,
                displayMsg:'{from}/{to} 共{total}条',
                onSelectPage:function(pageNumber, pageSize){
                    listNews(pageNumber,pageSize);
```

```
            }
        });
}
function checkUserName()  //登录前，校验用户信息
{
        var a=$('#uid').val();
        var b=$('#pwd').val();
        if(a==""){
            alert("请输入用户名");return;
            $.messager.alert('系统提示',"请输入用户名","warning");
            return;
        }
        if(b==""){$.messager.alert('系统提示',"请输入登录密码","warning");return;}
        $.ajax({
                url : base+"doLogin",
        //只封装和传输指定的数据
                data :{"uid":a,"pwd":b},
                type:"POST",
                success : function (res) {
                        if (res.ok) {
                                window.location.href=base+res.msg;
                        }else {$.messager.alert('系统提示',res.msg,"warning");          }
                        return false;
                },
                error : function(res) {$.messager.alert('系统提示',"系统错误！ ","error");        }
        });
}


//源代码，作用是显示动态效果
$(function(){
    airBalloon('div.air-balloon');
});
/*
@function  热气球移动
@update by julying , 2012/7/25
*/
function airBalloon(balloon){
    var viewSize = [] , viewWidth = 0 , viewHeight = 0 ;
    resize();
    $(balloon).each(function(){
        $(this).css({top: rand(40, viewHeight * 0.5 ) , left : rand( 10 , viewWidth - $(this).width() ) });
        fly(this);
    });
    $(window).resize(function(){
        resize()
```

```
            $(balloon).each(function(){
                $(this).stop().animate({top: rand(40, viewHeight * 0.5 ) , left : rand( 10 , viewWidth - 
$(this).width() ) } ,1000 , function(){
                    fly(this);
                });
            });
        });
        function resize(){
            viewSize = getViewSize();
            viewWidth = $(document).width() ;
            viewHeight = viewSize[1] ;
        }
        function fly(obj){
            var $obj = $(obj);
            var currentTop = parseInt($obj.css('top'));
            var currentLeft = parseInt($obj.css('left') );
            var targetLeft = rand( 10 , viewWidth - $obj.width() );
            var targetTop = rand(40, viewHeight /2 );
            /*求两点之间的距离*/
            var removing = Math.sqrt( Math.pow( targetLeft - currentLeft , 2 )  + Math.pow( targetTop - 
currentTop , 2 ) );
            /*每秒移动 24px ，计算所需要的时间，从而保持 气球的速度恒定*/
            var moveTime = removing / 24;
            $obj.animate({ top : targetTop , left : targetLeft} , moveTime * 1000 , function(){
                setTimeout(function(){
                    fly(obj);
                }, rand(1000, 3000) );
            });
        }
        function rand(mi,ma){
            var range = ma - mi;
            var out = mi + Math.round( Math.random() * range) ;
            return parseInt(out);
        }
        function getViewSize(){
            var de=document.documentElement;
            var db=document.body;
            var viewW=de.clientWidth==0 ?  db.clientWidth : de.clientWidth;
            var viewH=de.clientHeight==0 ?  db.clientHeight : de.clientHeight;
            return Array(viewW,viewH);
        }
    };
```

4.13.2 出错跳转页

当出现系统错误时，直接跳转到 nologin.jsp 页，页面代码如下。

```
<%@ page language="java" import="java.util.*" pageEncoding="UTF-8"%>
<%
String path = request.getContextPath();
String basePath = request.getScheme()+"://"+request.getServerName()+":"+request.getServerPort()+path
+"/";
%>

<!DOCTYPE HTML>
<html>
<head>
    <title>无效访问</title>
    <meta charset="UTF-8"/>
    <link href="<%=basePath%>include/img/logo.png" rel="icon"  type="image/x-icon" />
    <link  href='<%=basePath%>include/css/404.css?family=Love+Ya+Like+A+Sister'  rel='stylesheet'
type='text/css'>
    <script src="<%=basePath%>include/js/jquery.min.js" type="text/javascript" ></script>
</head>
<body>
  <div class="wrap">
    <div class="logo">
        <div class="errcode"><span>out</span></div>
        <p>对不起，您没有登录或者登录已超时!!!</p>
        <div class="sub">
          <p><a href="###"  id="index">返回首页</a></p>
        </div>
    </div>
  </div>
  <script type="text/javascript">
$("#index").click(function() { reDo(); });
setTimeout(reDo, 5000);
function reDo(){ top.location.href = "<%=basePath%>";   }
</script>
</body>
</html>
```

4.13.3　新闻阅读页

尝试在新闻阅读页 newsread.jsp 使用 Struts2 数据标签。页面中有如下几个特殊的地方：

- 引用标签库，定义前缀：<%@ taglib prefix="s" uri="/struts-tags" %>。
- 访问 Action 返回值中的对象属性，如新闻标题：<s:property value="news.title" />。
- 将 value 值当做 html 代码去解析，如新闻内容：<s:property value="news.content" escape="false"/>。
- 输出并格式化日期，如新闻发布时间：<s:date name="news.tjdate" format="yyyy-MM-dd"/>。

代码如下：

```
<%@ taglib prefix="s" uri="/struts-tags" %>
<!DOCTYPE html>
<html>
<head>
    <meta charset="UTF-8" />
    <title>读新闻</title>
    <link rel="shortcut icon" href="<%=basePath%>include/img/logo.png" type="image/x-icon" />
    <link rel="stylesheet" type="text/css" href="<%=basePath%>include/css/news.css">
    <script type="text/javascript" src="<%=basePath%>include/js/jquery.min.js"></script>
</head>
<body>
    <div style="background:#B3DFDA;padding:0 10px 0 10px;vertical-align: middle;">
        <img src="<%=basePath%>include/img/logo.png" width="126" height="50" />
        <div style="float:right;line-height:50px;margin-right:10px;font-size: 9pt;">
            <span>【</span><a style="color:blue;" href="javascript:window.close();"><span>
            关闭窗口</span></a><span>】</span>
        </div>
    </div>
<div class="ndetail">
    <div class="ntitle"><s:property value="news.title" /></div>
    <div class="nauthor">
        <div>来源:  <strong><s:property value="news.cruser" /></strong>   
        发布时间:    <strong><s:date name="news.tjdate" format="yyyy-MM-dd"/></strong>
          访问量:  <strong>[<span><s:property value="news.hitnum" /></span>]
        </strong></div>
    </div>
    <div class="nbody">
        <div id="vsb_content"> <s:property value="news.content"   escape="false"/></div>
    </div>
</div>
</body>
</html>
```

4.13.4 后台 Layout

新闻管理的后台布局页 admin.jsp，页面用 EasyUI Layout 实现布局，左侧动态加载树形菜单，当单击菜单项时，切换主窗口的 Tab 选项卡；用 swNewTab('新闻管理','<%=basePath%>goListNews");语句实现默认加载后台新闻列表 Tab；当用户单击页面右上角的“退出”按钮时，回到系统首页。

显示页面内容的代码如下：

```
<%@ page language="java" import="java.util.*" pageEncoding="UTF-8"%>
<%
String path = request.getContextPath();
String basePath = request.getScheme()+"://"+request.getServerName()+":"+request.getServerPort()+
path+"/";
%>
```

```
<!DOCTYPE html>
<html>
  <head>
    <title>后台管理系统</title>
    <meta charset="UTF-8" />
        <link rel="shortcut icon" href="<%=basePath%>include/img/logo.png">
        <link rel="stylesheet" type="text/css" href="<%=basePath%>include/easyui/themes/default/
            easyui.css">
        <link rel="stylesheet" type="text/css" href="<%=basePath%>include/easyui/themes/icon.css">
        <script type="text/javascript" src="<%=basePath%>include/js/jquery.min.js"></script>
        <script type="text/javascript" src="<%=basePath%>include/easyui/jquery.easyui.min.js">
            </script>
        <script type="text/javascript" src="<%=basePath%>include/easyui/locale/easyui-lang-
            zh_CN.js"></script>
        <script type="text/javascript" src="<%=basePath%>include/js/lrwtab.js"></script>
  </head>
 <body class="easyui-layout">
      <div data-options="region:'north',border:false"
           style="background:#fff;padding:0 10px 0 10px;vertical-align: middle;">
        <img src="<%=basePath%>include/img/logo.png" width="126" height="50" />
        <div style="float:right;line-height:50px;margin-right:10px;">
              <a id="logout" href="#" class="easyui-linkbutton" data-options=
                  "iconCls:'icon-cancel'">退出</a>
        </div>
        <div style="float:right;line-height:50px;margin-right:10px;">登录用户：${me.xm} |</div>
      </div>
      <div data-options="region:'west',split:true,title:'系统菜单'" style="width:180px;padding:10px;">
          <ul id="menutree" class="easyui-tree"></ul>
      </div>
      <div data-options="region:'south',border:false" style="height:30px;padding:5px;text-align:center;
          font-family: arial;
      color: #A0B1BB;">Copyright © 2017 JavaEE. All Rights Reserved.
      </div>
      <div data-options="region:'center'">
          <div id="tabs" class="easyui-tabs" fit="true" border="false">
          </div>
      </div>
</body>
<script>
var opened_node;
$(function(){
$("#menutree").tree(
{
    url : "<%=basePath%>menu",
    animate : true,
    onClick : function(node) {
```

```
                if (!node.attributes) {
                        if (!opened_node) {
                                $("#menutree").tree('expand', node.target);
                                opened_node = node.target;
                        } else if (opened_node != node.target) {
                                $("#menutree").tree('collapse', opened_node);
                                $("#menutree").tree('expand', node.target);
                                opened_node = node.target;
                        }
                } else {
                    swNewTab(node.text,"<%=basePath%>" +node.attributes.url);
                }
            },
            onLoadSuccess : function(node, data) {
                $("#menutree").tree('expandAll');
            }
    });
    $("#logout").click(function(){
        top.location.href="<%=basePath%>doLogout";
    });
    swNewTab('新闻管理',"<%=basePath%>goListNews");
    });
    </script>
    </html>
```

4.13.5 新闻列表页

后台新闻列表页 newslist.jsp，页面上使用 datagrid 加载分页新闻，每条新闻后有修改和删除的链接，工具栏中有标题关键字输入框、查询按钮。

- 新闻标题过长时，能够自动折断。
- 新闻日期显示格式为 yyyy-MM-dd hh:mm。
- 单击“修改”按钮，会跳转到修改新闻的窗口。
- 单击“删除”按钮，会弹出确认对话框。
- 输入标题关键词，单击“查询”按钮，会显示新闻标题与关键词匹配的新闻。

显示页面内容的代码如下：

```
<%@ page language="java" import="java.util.*" pageEncoding="UTF-8"%>
<%
String path = request.getContextPath();
String basePath = request.getScheme()+"://"+request.getServerName()+":"+ request.getServerPort()+
path+"/";
%>
<!DOCTYPE html>
<html>
  <head>
    <title>新闻列表</title>
```

```
        <meta charset="UTF-8" />
            <link rel="shortcut icon" href="<%=basePath%>include/img/logo.png">
            <link rel="stylesheet" type="text/css" href="<%=basePath%>include/easyui/themes/default/
                easyui.css">
            <link rel="stylesheet" type="text/css" href="<%=basePath%>include/easyui/themes/icon.css">
            <script type="text/javascript" src="<%=basePath%>include/js/jquery.min.js"></script>
            <script type="text/javascript" src="<%=basePath%>include/easyui/jquery.easyui.min.js">
                </script>
            <script type="text/javascript" src="<%=basePath%>include/easyui/locale/easyui-lang-
            zh_CN.js"></script>
    </head>
<body>
  <table id="dg" cellpadding="2" ></table>
<div id="tb" style="padding:5px;">
        <input id="s_name" class="easyui-textbox"   data-options="prompt:'标题关键字...'" style=
            "width:200px;height:32px">
        <a id="s_news" href="#" class="easyui-linkbutton" data-options="iconCls:'icon-search'">查询</a>
</div>
</body>
<script>
var s_name="",id="",title="";
function loadGrid(){
        s_name=$("#s_name").val();
        $("#dg").datagrid({
                width:800,height:500,nowrap:false,
                striped:true,border:true,collapsible:false,
                url:"<%=basePath%>listNews",
                queryParams:{"s_name":s_name},
                pagination:true,
                rownumbers:true,
                fitColumns:true,pageSize:20,
                loadMsg:'数据加载中...' ,
                columns:[[
                        {title:'标题', field:'title',width:200,formatter: function(value,row,index){
                            return '<span style="white-space: nowrap;" title='+value+'>'+
                                (value?value:'')+'</span>';
                        }},
                        {title:'发布时间', field:'tjdate',width:100, formatter: function(value,row,index){
                              return (new Date(row.tjdate).Format("yyyy-MM-dd hh:mm"));
                          }},
                        {title:'操作', field:'hitnum',width:100, formatter: function(value,row,index){
                            var p="<a href=\"javascript:editNews('"+row.id+"')\">修改</a>";
                            p+="| <a href=\"javascript:delNews('"+row.id+"','"+row.title+"')\">删除</a>";
                            return p;
                         }}
                    ]],
```

```
                    toolbar:'#tb'
            });
        }
        function editNews(id){
            parent.swNewTab("修改新闻信息","<%=basePath%>goEditNews?id="+id);
        }
        function delNews(newsid,title0){
            id=newsid;title=title0;
            parent.$.messager.confirm("系统提示", "您确认要删除“"+title+"”吗？", function(r){
              if (r){
                    $.ajax({
                        url:"<%=basePath%>doDelNews",
                        data:{"id":id},
                        type:"post",
                        success: function(res){
                            if(res.delflag){
                                parent.$.messager.alert("系统提示","您已删除新闻："+title,"info");
                                id="";s_name="";
                                loadGrid();
                            }else {
                                parent.$.messager.alert("系统提示",res,"error");
                            }
                            return false;
                        },
                        error:function(res){
                            parent.$.messager.alert("系统提示","系统错误","error");
                        }
                    })
                }
            });
        }
        $(function(){
            loadGrid();
            $("#s_news").click(function(){
                s_name=$("#s_name").val();
                loadGrid();
            });
            $("#tb").bind("keydown",function(e){
                var theEvent = e || window.event; // 兼容 FF 和 IE 和 Opera
                var code = theEvent.keyCode || theEvent.which || theEvent.charCode;
                if (code == 13) {
                    $("#s_news").click();
                }
            });
        })
        </script>
        </html>
```

4.13.6　新闻添加页

后台新闻添加页 newsadd.jsp，页面上使用百度在线编辑器 UEditor 实现图文混排、文件上传、发布新闻。

- 直接使用百度提供的实现上传的控制器 controller.jsp，如图 4-29 所示安装 UEditor 资源，特别注意图中方框标记的文件。

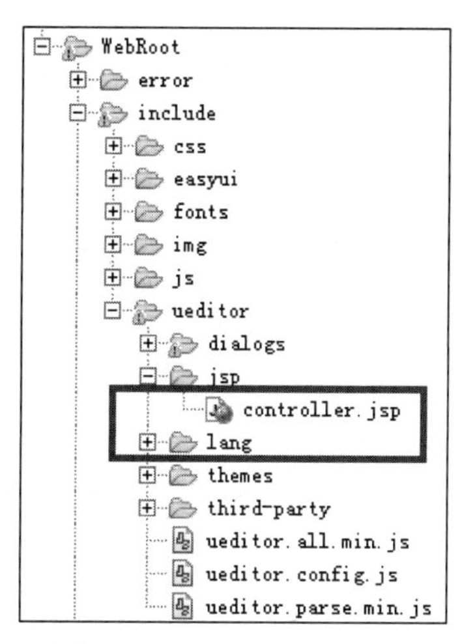

图 4-29　安装 UEditor 资源

- 由于 Struts 默认会拦截上传功能，一方面在后端和配置文件中修改了过滤器，另一方面修改前端的请求。修改 ueditor.config.js，使用官方的 JSP 页面响应上传请求。
  serverUrl: URL + "jsp/controller.jsp"
- 把 UEditor 中的 config.json 文件放在当前项目的 config 目录下，然后修改 config.json 文件，把原始文件中所有默认的上传保存路径/ueditor/jsp/upload/全部修改为当前项目设计的路径前缀/upload/。例如，图片的上传保存路径修改为：
  "imagePathFormat": "/upload/image/{yyyy}{mm}{dd}/{time}{rand:6}"。

显示页面 newsadd.jsp 内容的代码如下：

```
<%@ page language="java" import="java.util.*" pageEncoding="UTF-8"%>
<%
String path = request.getContextPath();
String basePath = request.getScheme()+"://"+request.getServerName()+":"+request.getServerPort()+
path+"/";
%>
<!DOCTYPE html>
<html>
<head>
    <title>添加新闻</title>
    <meta charset="UTF-8"/>
    <link rel="shortcut icon" href="<%=basePath%>include/img/logo.png">
    <link rel="stylesheet" type="text/css" href="<%=basePath%>include/easyui/themes/default/easyui.css">
```

```
    <link rel="stylesheet" type="text/css" href="<%=basePath%>include/easyui/themes/icon.css">
    <script type="text/javascript" src="<%=basePath%>include/js/jquery.min.js"></script>
    <script type="text/javascript" src="<%=basePath%>include/easyui/jquery.easyui.min.js"></script>
    <script type="text/javascript" src="<%=basePath%>include/easyui/locale/easyui-lang-zh_CN.js">
        </script>
    <script>var base="<%=basePath%>";</script>
    <script type="text/javascript" charset="UTF-8" src="<%=basePath%>include/ueditor/
        ueditor.config.js"></script>
    <script type="text/javascript" charset="UTF-8" src="<%=basePath%>include/ueditor/
        ueditor.all.min.js"> </script>
    <script type="text/javascript" charset="UTF-8" src="<%=basePath%>include/ueditor/lang/zh-cn/
        zh-cn.js"></script>
</head>
<body>
        <div class="easyui-panel" style="padding:5px 2px">
        <form>
                <table cellpadding="5">
                <tr><td style="width:120px;">新闻标题：</td><td style="width:880px;">
                        <input id="title" class="easyui-textbox" data-options="prompt:'新闻标题',
                            required:true" style="width:90%;height:32px">
                </td></tr>
                <tr><td>新闻发布者：</td><td>
                        <input id="cruser" class="easyui-textbox" value="${me.xm} " data-options=
                            "prompt:'发布人',required:true" style="width:90%;height:32px">
                </td></tr>
                <tr><td style="vertical-align: top;">新闻内容：</td><td>
                        <script id="content" type="text/plain" style="width:89%;height:300px;"></script>
                </td></tr>
                </table>
                </form>
                <div style="text-align:center;">
                        <a id="savenews" href="#" class="easyui-linkbutton" iconCls="icon-ok" style=
                            "width:132px;height:32px">保存</a>
                </div>
        </div>
  <script type="text/javascript">
  var ue;
  $(function(){
      ue = UE.getEditor('content');
      $('#savenews').click(function(){   //发布新闻前，要校验
          var a=$("#title").textbox("getValue");
          var b=ue.getContent();
          var c=$("#cruser").textbox("getValue");
              if(a.length<=0){$.messager.alert("系统提示","必须填写新闻标题","warning");return;}
              if(b.length<=0){$.messager.alert("系统提示","必须填写新闻内容","warning");return;}
```

```
            if(c.length<=0){$.messager.alert("系统提示","必须填写发布人姓名或者发布机构
                名称","warning");return;}
            $.ajax({
                type: 'POST',
                url : "<%=basePath%>saveAddNews",
                data : {"news.title":a,"news.content":b,"news.cruser":c},
                success : function (res) {
                    if(res.ok){
                        parent.$.messager.alert("系统提示","你已添加新闻:"+$("#title").
                            val(),"info");
                    }else{
                        parent.$.messager.alert("系统提示","添加失败！","error");
                    }
                    return false;
                },
                error : function(res) {parent.$.messager.alert("系统提示","系统错误！","error");}
            });
        });
});
</script>
</body>
</html>
```

注意：如果 UEditor 无法实现图片或其他文件上传，需要仔细检查以下 5 个方面。

（1）UEditor 需要 6 个 Jar 包，尤其是 commons-lang3-3.2.1.jar。

（2）\include\ueditor 目录结构，尤其注意\include\ueditor\jsp\controller.jsp。

（3）config.json 的存放路径，文件中 PathFormat 的正确配置。

（4）MyStrutsFilter.java。

（5）web.xml。

4.13.7　新闻修改页

后台新闻修改页 newsedit.jsp，新闻修改页内容与新闻添加页差不多，不同的地方是：

- 需要加载指定 id 的新闻的多项内容到页面。
- 修改完成后，请求保存的 URL；封装的信息除新闻标题、新闻内容、新闻发布者外，增加了新闻的 id。

显示页面 newsedit.jsp 的内容代码如下：

```
<%@ page language="java" import="java.util.*" pageEncoding="UTF-8"%>
<%
String path = request.getContextPath();
String basePath = request.getScheme()+"://"+request.getServerName()+":"+request.getServerPort()+
path+"/";
%>
<!DOCTYPE html>
<html>
```

```
<head>
    <meta charset="UTF-8" />
    <title>修改新闻</title>
    <link rel="shortcut icon" href="<%=basePath%>include/img/logo.png">
    <link rel="stylesheet" type="text/css" href="<%=basePath%>include/easyui/themes/default/
        easyui.css">
    <link rel="stylesheet" type="text/css" href="<%=basePath%>include/easyui/themes/icon.css">
    <script type="text/javascript" src="<%=basePath%>include/js/jquery.min.js"></script>
    <script type="text/javascript" src="<%=basePath%>include/easyui/jquery.easyui.min.js"></script>
    <script type="text/javascript" src="<%=basePath%>include/easyui/locale/easyui-lang-zh_CN.js">
        </script>
    <script>var base="<%=basePath%>";</script>
    <script type="text/javascript" charset="UTF-8" src="<%=basePath%>include/ueditor
        /ueditor.config.js"></script>
    <script type="text/javascript" charset="UTF-8" src="<%=basePath%>include/ueditor/
        ueditor.all.min.js"> </script>
    <script type="text/javascript" charset="UTF-8" src="<%=basePath%>include/ueditor/lang/
        zh-cn/zh-cn.js"></script>
</head>
<body>
      <div class="easyui-panel" style="padding:5px 2px">
      <form>
            <table cellpadding="5">
            <tr><td style="width:100px;">新闻标题：</td><td style="width:880px;">
                  <input id="title" class="easyui-textbox" data-options="prompt:'新闻标题',
                        required:true" style="width:90%;height:32px">
            </td></tr>
            <tr><td>新闻发布者：</td><td>
                  <input id="cruser" class="easyui-textbox" data-options="prompt:'发布人',
                       required:true" style="width:90%;height:32px">
            </td></tr>
            <tr><td style="vertical-align: top;">新闻内容：</td><td>
                  <script id="content" type="text/plain" style="width:89%;height:300px;"></script>
            </td></tr>
            </table>
      </form>
      <div style="text-align:center;">
            <a id="savenews" href="#" class="easyui-linkbutton" iconCls="icon-ok" style=
                "width:132px;height:32px">保存</a>
      </div>
      </div>
<script type="text/javascript">
var ue;
$(function(){
    ue = UE.getEditor('content');
```

```
        $("#title").textbox("setValue","${news.title}");
        ue.ready(function() {
            ue.setContent("");
            ue.execCommand('insertHtml', '${news.content}');
        });
        $("#cruser").textbox("setValue","${news.cruser}");
        $('#savenews').click(function(){   //发布新闻前，要校验
            var a=$("#title").textbox("getValue");
            var b=ue.getContent();
            var c=$("#cruser").textbox("getValue");
                if(a.length<=0){$.messager.alert("系统提示","必须填写新闻标题","warning");return;}
                if(b.length<=0){$.messager.alert("系统提示","必须填写新闻内容","warning");return;}
                if(c.length<=0){$.messager.alert("系统提示","必须填写发布人姓名或者发布机构名称
                    ","warning");return;}
                $.ajax({
                        type: 'POST',
                        url : "<%=basePath%>saveEditNews",
                        data : {"news.title":a,"news.content":b,"news.cruser":c,"news.id":${news.id}},
                        success : function (res) {
                                if(res.ok){
                                        parent.$.messager.alert("系统提示","你已修改新闻:"+
                                            $("#title").val(),"info");
                                }else{
                                        parent.$.messager.alert("系统提示","修改失败！","error");
                                }
                                return false;
                        },
                        error : function(res) {parent.$.messager.alert("系统提示","系统错误！","error");}
                });
        });
});
</script>
</body>
</html>
```

4.14　增强安全

可以在 Utils 包中设置一个自定义的过滤器，如 Loginfilter，实现对某些请求不过滤，直接通过。某些请求必须过滤，过滤检测不通过（拦截）时跳转到指定页面。例如，当未登录用户访问后台页面时，自动跳转到 nologin.jsp 页面。

Filter 是服务器端的组件，用来过滤 Web 请求。当一个 Web 请求时，Web 容器会先检查请求的 URL 是否设置了 Filter，如果已设置则执行该 Filter 的 doFilter 方法。所有 Filter 都实现了 javax.servlet.Filter 接口，doFilter 是定义在该接口中的最重要的方法。过滤器有先后顺序，

当有多个过滤器时，称之为过滤器链，形如：

request -> filter1 -> filter2 ->filter3 -> …. -> request resource

当执行 chain.doFilter(request,response)时，将请求转发给过滤器链上的下一个 Filter 对象，如果没有 Filter 了，那就是请求的资源了。

4.14.1 过滤器 LoginFilter

自定义的 LoginFilter 过滤器与 web.xml 有关，如果前端请求 URL 包含 excludeStrings 参数中的一个字符串或者请求系统根目录“/”时，直接将请求转发下一个过滤器；否则，查看用户是否已经登录，如果用户没有登录，则由 wrapper.sendRedirect(redirectPath)语句实现跳转到 redirectPath 参数指定的页面。

```
public class LoginFilter extends HttpServlet implements Filter {
    public FilterConfig config;
    public boolean isContains(String container, String[] regx) {
        boolean result = false;
        for (int i = 0; i < regx.length; i++) {
            if (container.indexOf(regx[i]) != -1) {
                return true;
            }
        }
        return result;
    }
    @Override
    public void doFilter(ServletRequest request, ServletResponse response,
            FilterChain chain) throws IOException, ServletException {
        HttpServletRequest hrequest = (HttpServletRequest)request;
        String ctx=hrequest.getContextPath();      //newsssh
        String uri=hrequest.getRequestURI();
        HttpServletResponseWrapper wrapper = new HttpServletResponseWrapper
            ((HttpServletResponse) response);
        String redirectPath = ctx + config.getInitParameter("redirectPath");
        String excludeStrings = config.getInitParameter("excludeStrings");
        String[] excludeList = excludeStrings.split(";");
        if (this.isContains(uri, excludeList) || uri.equals(ctx+"/")) {
            chain.doFilter(request, response);
            return;
        }
        User user = ( User ) hrequest.getSession().getAttribute("me");
        if (user == null ) {
            System.out.println("成功拦截到非法用户企图入侵网站后台    :  " + uri);
            wrapper.sendRedirect(redirectPath);
            return;
        }
        chain.doFilter(request, response);
```

```
    }
    @Override
    public void init(FilterConfig config) throws ServletException {
        this.config = config;
    }
}
```

4.14.2 配置 LoginFilter

在 web.xml 中添加 LoginFilter 过滤器的配置。设置 excludeStrings 参数，参数值 param-value 为不需要被过滤器拦截的字符串；设置 redirectPath 参数，参数值 param-value 为被拦截后指定的重定向页。

过滤器 LoginFilter 的配置内容如下：

```
<filter>
    <filter-name>loginFilter</filter-name>
    <filter-class>cn.lrw.newsssh.utils.LoginFilter</filter-class>
    <init-param>
        <param-name>excludeStrings</param-name>
        <param-value>/include;doLogin;/error;listNews;getNews;getCountNews;upload;.js;.css;.jpg;
            .png;.gif;.ico</param-value>
    </init-param>
    <init-param>
        <param-name>redirectPath</param-name>
        <param-value>/error/nologin.jsp</param-value>
    </init-param>
</filter>
<filter-mapping>
    <filter-name>loginFilter</filter-name>
    <url-pattern>/*</url-pattern>
</filter-mapping>
```

当完成项目搭建、后端开发、前端开发、系统配置等所有工作后，再运行项目，测试各项功能。如果发现有问题，则进行调试，解决存在的各种问题。

4.15　考核任务

正确实现基于 SSH 框架的新闻信息增删改查，添加与修改新闻可以上传图片等文件。

（1）添加新闻。（20 分）

（2）修改新闻。（20 分）

（3）后台新闻列表。（20 分）

（4）删除新闻。（20 分）

（5）前台新闻列表。（10 分）

（6）阅读新闻。（10 分）

本章小结

本章开发基于 SSH 框架的项目，采用前端与后端分离开发的方法。该方法适合团队协作，在一定程度上可以并行，缩短开发时间，但需要开发人员先协商，约定前后端有关联的内容，例如请求的路径、方法的命名、参数的数量类型等。也需要测试人员对前后端的开发技术都要掌握一些。

基于 SSH 框架的项目的工作原理是：

（1）当在浏览器中有请求 http://localhost:8080/newsssh/doLogin 时，就会向服务器端（Tomcat）发送一个请求。

（2）Tomcat 会解析 URL，从中找到 Web 应用项目 newsssh 的 web.xml 文件。

（3）web.xml 中<url-pattern>/*</url-pattern>会过滤掉所有地址，然后在 Filter 中定义的 filter-class 会处理 URL 中的 doLogin 请求。

（4）StrutsPrepareAndExecuteFilter 接收到 URL 之后，首先查询 struts.xml 中的 namespace，然后将 URL 中 namespace 值后面的路径读取到。

（5）继续在 Struts 的 action 标签中查找是否有 doLogin 这个 name，如果有且发现 action 中有 class 属性，则会执行 class 里面的 method 属性指定的方法。

（6）执行方法后，返回一个 String 字符串或者 JSON 对象。根据返回值（result 中的值），重定向、转发或在 JS 中分析处理，如果有错则会重定向到 error 中的 JSP 页面，否则通过 JSP 反馈到客户端浏览器。

第5章　基于 SSM 的项目实战

本章介绍采用前端与后端分离开发的方式来开发基于 SSM 的项目。首先采用向导搭建 SSM 框架的 Web 项目（主要是引入 Spring 和 MyBatis 框架的 Jar 包，自动配置一些参数）；然后编写后端的 Java 类：entity、DAO、Service、Web，配置 MyBatis、Spring，创建 Mapper 映射文件；最后参照第 3 章的前端页面快速开发当前项目的前端页面。

5.1　SSM 简介

SSM（Spring+SpringMVC+MyBatis）框架，由 Spring（内含 SpringMVC）、MyBatis 两个开源框架整合而成，常作为数据源较简单的 Web 项目的框架。

Spring 管理各层的组件。Spring 的 IOC 容器可以装载 Bean，不用在每次使用 Bean 类的时候为它初始化，很少看到关键字 new。

SpringMVC 作为控制器（Controller），其工作原理如图 5-1 所示。

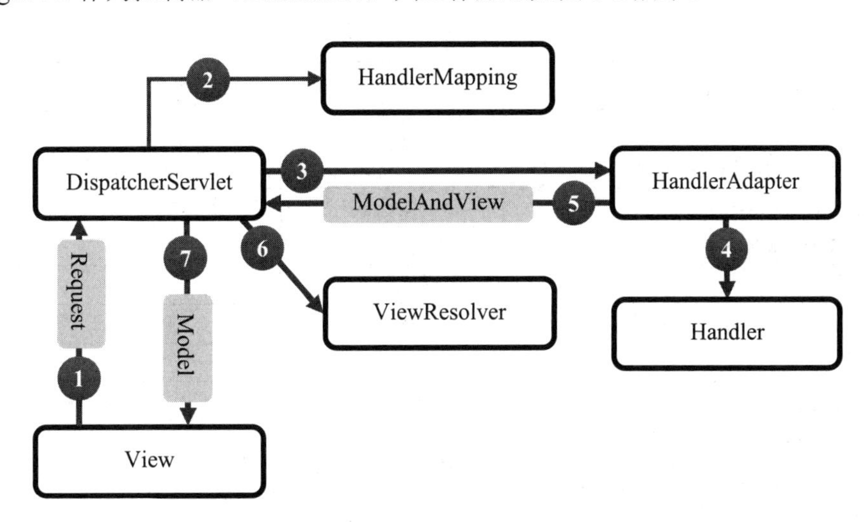

图 5-1　SpringMVC 工作原理

（1）客户端发出一个 HTTP 请求给 Web 服务器，Web 服务器对 HTTP 请求进行解析，如果匹配 DispatcherServlet 的请求映射路径（在 web.xml 中指定），则 Web 容器将请求转交给 DispatcherServlet。

（2）DipatcherServlet 接收到这个请求之后将根据请求的信息（包括 URL、HTTP 方法、请求报文头和请求参数 Cookie 等）以及 HandlerMapping 的配置找到处理请求的处理器（Handler）。

（3）～（4）DispatcherServlet 根据 HandlerMapping 找到对应的 Handler，将处理权交给 Handler（Handler 将具体的处理进行封装），再由具体的 HandlerAdapter 对 Handler 进行具体的调用。

（5）Handler 对数据处理完成以后将返回一个 ModelAndView 对象给 DispatcherServlet。

（6）Handler 返回的 ModelAndView 只是一个逻辑视图，并不是一个正式的视图，DispatcherSevlet 通过 ViewResolver 将逻辑视图转化为真正的视图 View。

（7）Dispatcher 通过 Model 解析出 ModelAndView 中的参数，将解析出最终完整的 View 返回给客户端。

MyBatis 负责持久化层。通过 SessionFactoryBuider 由 XML 配置文件生成 SessionFactory；然后由 SessionFactory 生成 Session；最后由 Session 来开启执行事务和 SQL 语句。与 Hibernate 比较，MyBatis 能自由控制 SQL，可以进行更为细致的 SQL 优化，可以减少查询字段。

5.2 创建 SSM 项目

创建项目前，先设计好项目工程结构，准备好所需的 Jar 包、JS 库、CSS 库、MySQL 数据库等资源。

5.2.1 项目工程结构

当前项目的工程结构设计见表 5-1。

表 5-1 项目工程结构

newsssm	项目名称
├─src	源码文件夹
│ ├─cn.lrw.newsssm.web	控制器，SpringMVC 在这里发挥作用
│ ├─cn.lrw.newsssm.entity	存放数据库表/视图对应的实体类
│ ├─cn.lrw.newsssm.dao	DAO 类，封装对数据访问的方法
│ ├─cn.lrw.newsssm.service	业务逻辑类
│ └─cn.lrw.newsssm.utils	公用方法类
├─reources	
│ ├─mapper	存放 MyBatis 实体类映射文件
│ ├─applicationContext.xml	配置 Spring、MyBatis
│ ├─log4j.properties	系统日志配置
│ ├─config.json	UEditor 的配置文件
│ └─mybatis-config.xml	配置 Mybatis 核心文件
└─WebRoot	Web 项目根目录
├─error	存放异常访问提示页，如 nologin.jsp、403.jsp、404.jsp 等
├─include	分类存放网页中引用的 JS、CSS、图像类文件
│ ├─css	
│ ├─js	
│ ├─img	
│ ├─easyui	EasyUI 框架
│ └─ueditor	百度的可视化 HTML 编辑器

续表

Newsssm	项目名称
├─upload	存放上传的文件
├─WEB-INF	Java Web 应用的安全目录
│　├─lib	存放项目需要的 Jar 包
│　├─web	存放网站中网页文件，如 JSP 等
│　└─web.xml	Web 工程的配置文件
└─index.jsp	Web 工程默认的首页文件

5.2.2　准备 Jar 包和 JS 库

本项目使用表 5-2 所示的的 Jar 包，如果需要其他的 Jar 包，可以从 Maven 查找和下载。

表 5-2　项目所需 Jar 包

Jar	说明
mysql-connector-java	MySQL 数据库驱动
druid	数据库连接池 Druid，带强大的监控功能
mybatis(2)	一个基于 Java 的持久层框架，映射接口和 POJO
Gson	google 开发的 Java API，用于转换 Java 对象和 JSON 对象
Log4j	用 Java 编写的可靠，快速和灵活的日志框架
shiro	用于认证、授权、加密、会话管理、与 Web 集成、缓存等
ueditor(6)	百度在线编辑器。依赖 json.jar、commons-codec-1.10.jar、commons-io-2.4.jar、commons-fileupload-1.3.3.jar、commons-lang3-3.2.1.jar
Spring 所需的 Jar	通过 MyEclipse 向导添加

本项目还将使用到一些常用的 JS 库文件，如 jQuery，EasyUI，UEditor 等。

5.2.3　新建 Web 项目

（1）新建 Web Project，然后设置基本参数，如图 5-2 所示。

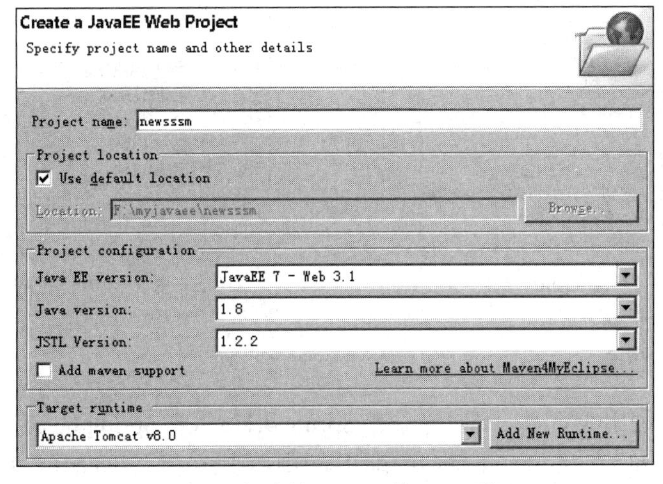

图 5-2　新建基于 SSM 的 Web 项目

（2）如图 5-3 所示，添加源文件夹 resources，用于存放配置文件。

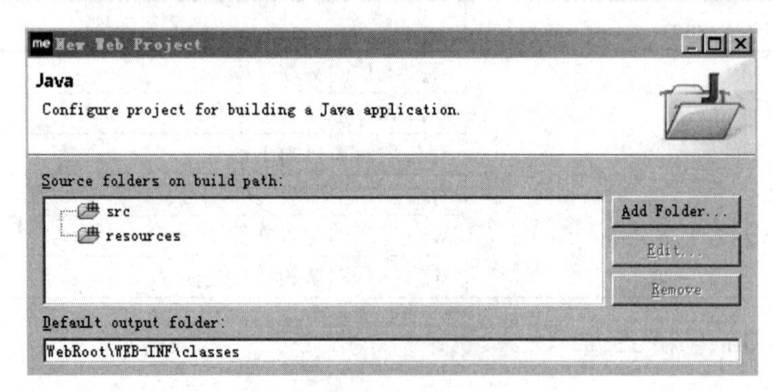

图 5-3　添加源文件夹 resources

（3）在下一步窗口中，勾选生成 index.jsp、web.xml，然后单击 Finish 按钮，完成创建。

（4）把按照表 5-2 准备的 Jar 包放进项目的 lib 文件夹，如图 5-4 所示。

（5）运行测试，确保无报错并正确显示默认的首页。然后参照“项目工程结构”创建包（package）或者文件夹，如图 5-5 所示。

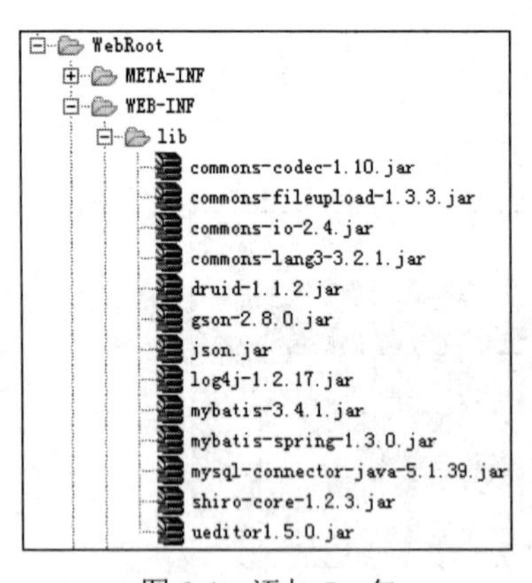

图 5-4　添加 Jar 包

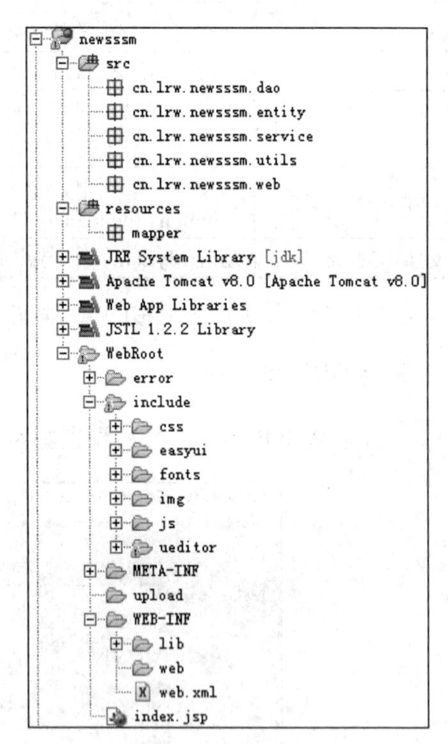

图 5-5　项目工程结构

5.2.4　添加 Spring

参照 4.2.5 节，利用 MyEclipse 向导添加 Spring 4.1。完成 Spring 的安装后，如果配置文件 applicationContext.xml 不在 resources 目录下，则将其移到 resources 目录下。配置文件 applicationContext.xml 中已经自动添加了 Spring 监听器 listener。

5.2.5　添加数据源

假定当前项目所需的数据库 dbnews3（参照 2.3 节）已经在 MySQL 中创建完成。数据库 dbnews3 中有 3 张表：user、news、cmenu，手工输入 cmenu 表的数据，如图 5-6 所示。

dbnews3.cmenu: 3 rows total (approximately)

id	pid	name	url	permission
1	0	新闻管理	(NULL)	1
2	1	新闻列表	news/goList	1
3	1	添加新闻	news/goAdd	1

图 5-6　cmenu 表的数据

5.2.6　创建 entity 类

在 cn.lrw.newsssm.entity 包中创建数据库 3 张表的 entity 类，创建 EasyUITree 形菜单数据的 entity 类，如图 5-7 所示。

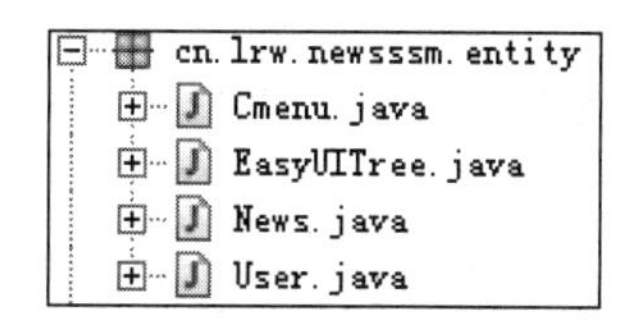

图 5-7　cn.lrw.newsssm.entity 包中的 entity 类

entity 类 Cmenu、EasyUITree、News 和 User 的形式如下，其中 EasyUITree 类的创建参照 4.5 节封装 Tree 型数据。

```
public class Cmenu  {
    private Integer id;
    private Integer pid;
    private String name;
    private String url;
    private String permission;
// 省略构造方法，getter 和 setter
}

public class User  {
    private String uid;
    private String xm;
    private String pwd;
    private String bj;
    private String role;
// 省略构造方法，getter 和 setter
}

public class News  {
    private Integer id;
```

```java
    private String title;
    private String content;
    private Date tjdate;
    private String cruser;
    private Integer hitnum;
// 省略构造方法，getter 和 setter
}
public class EasyUITree {
    private String id;
    private String text;
    private Boolean checked = false;
    private Map<String,Object> attributes;
    private String state = "closed";
public List<EasyUITree> children;
// 省略构造方法，getter 和 setter
}
```

5.2.7 配置 dataSource

在 applicationContext.xml 中添加数据库连接池配置，使用了阿里巴巴开源平台上的 druid，内容如下：

```xml
<bean id="dataSource" class="com.alibaba.druid.pool.DruidDataSource" destroy-method="close">
    <!-- 配置数据库相关参数 properties 的属性 -->
    <property name="driverClassName" value="com.mysql.jdbc.Driver" />
    <property name="url" value="jdbc:mysql://127.0.0.1:3066/dbnews3?useUnicode=true&amp;
        characterEncoding=UTF-8" />
    <property name="username" value="root" />
    <property name="password" value="" />
</bean>
```

5.2.8 配置 SpringMVC

在 web.xml 中，主要配置 SpringMVC 的核心控制器 DispatcherServlet 和防止 post 中文乱码的 Filter。SprintMVC 的 url-pattern 不能随意配置，它的作用是对请求的后缀进行筛选，默认匹配所有的请求。如果配置“*.do”，则拦截所有带后缀 do 的请求，对静态资源放行；如果配置“/”，则拦截所有带.jsp 的请求，同时对静态资源拦截；如果配置“/*”，则所有请求可以到达 Action 中，但转到 JSP 时再次被拦截，不能访问到 JSP。

```xml
<servlet>
    <servlet-name>SpringMVC</servlet-name>
    <servlet-class>org.springframework.web.servlet.DispatcherServlet</servlet-class>
    <init-param>
        <param-name>contextConfigLocation</param-name>
        <param-value>classpath:applicationContext.xml</param-value>
    </init-param>
</servlet>
```

```
<servlet-mapping>
    <servlet-name>SpringMVC</servlet-name>
    <url-pattern>/</url-pattern>
</servlet-mapping>

<!-- 解决中文乱码 -->
<filter>
    <filter-name>encodingFilter</filter-name>
    <filter-class>org.springframework.web.filter.CharacterEncodingFilter</filter-class>
    <init-param>
        <param-name>encoding</param-name>
        <param-value>UTF-8</param-value>
    </init-param>
</filter>
<filter-mapping>
    <filter-name>encodingFilter</filter-name>
    <url-pattern>/*</url-pattern>
</filter-mapping>
```

5.2.9　运行项目

运行基于 SSM 框架的 JavaEE 项目，如果控制台 Console 没有报错信息，则在浏览器可以看到正常的首页信息。

5.2.10　清理 Jar 包

通过向导添加 Spring 框架，引用了少量不需要的 Jar 包，可以在清理后重新引用，具体方法参照 4.2.11，最后删除 Spring 的 Library，重新引用清理后放进 lib 中的 Jar 包。通过向导添加的 Spring 相关的包，都是同一版本，减少了冲突的可能性，图 5-8 所示是清理后剩余的所需的 Jar 包。

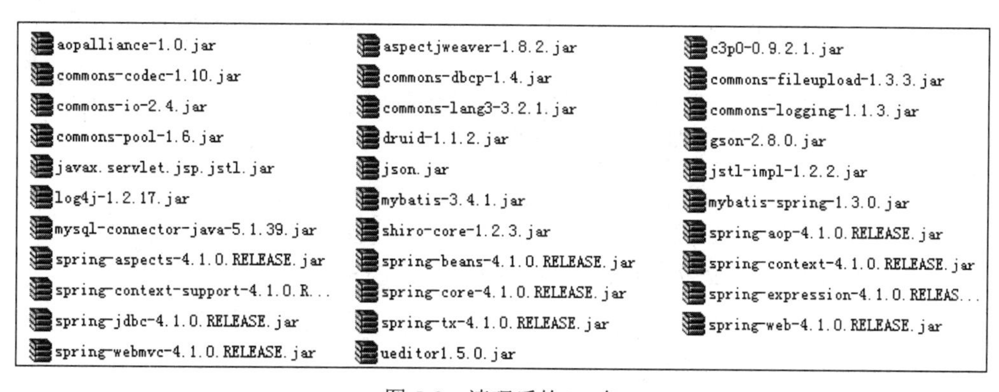

图 5-8　清理后的 Jar 包

重新运行当前项目，查看是否运行正常、是否可以正常显示项目首页。

5.3 考核任务

正确实现 SSM 项目的搭建。检查点：

（1）可以在浏览器看到默认首页 index.jsp 的内容，无乱码。（40 分）

（2）MyEclipse 的 Console 中没有报错（如 Exception）信息。（30 分）

（3）项目中已添加 Spring 框架、已配置 SpringMVC。（30 分）

5.4 日志系统

在 resources 中添加 log4j.properties，配置内容与 3.4 节中相同，以便在调试时看到更多的日志。

5.5 配置 Spring+Mybatis

简单配置 mybatis-config.xml；在 applicationContext.xml 中配置 Spring-dao、Spring-service、Spring-web 相关参数，配置过程中使用 tx、context 和 mvc 前缀，需要添加 XSD 的支持。以 MyEclipse Spring Config Editor 专用编辑器打开 Spring 配置文件 applicationContext.xml，切换到 Namespaces，勾选 tx、context 和 mvc 所对应的 XSD 命名空间，实现添加。

5.5.1 配置 MyBatis

在 resources 文件夹里新建 mybatis-config.xml 文件。主要配置：使用自增主键；使用列别名；开启驼峰命名转换，如 create_time -> createTime。

```
<?xml version="1.0" encoding="UTF-8"?>
<!DOCTYPE configuration PUBLIC "-//mybatis.org//DTD Config 3.0//EN" "http://mybatis.org/
dtd/mybatis-3-config.dtd">
<configuration>
  <!-- 配置全局属性 -->
  <settings>
    <!-- 使用 jdbc 的 getGeneratedKeys 获取数据库自增主键值 -->
    <setting name="useGeneratedKeys" value="true" />
    <!-- 使用列别名替换列名 默认:true -->
    <setting name="useColumnLabel" value="true" />
    <!-- 开启驼峰命名转换:Table{create_time} -> Entity{createTime} -->
    <setting name="mapUnderscoreToCamelCase" value="true" />
  </settings>
</configuration>
```

5.5.2 配置 Spring-dao

与 5.2.7 节的配置相同，在 DAO 层配置数据库连接池 dataSource。然后再进行以下配置：

（1）配置激活@Required、@Autowired、@Resource 等注解，让已注册的 Bean 开始工作。

```
<context:annotation-config />
```

（2）配置 SqlSessionFactory 对象，Spring 和 MyBatis 整合完美，不需要配置 MyBatis 的映射文件。

```
<bean id="sqlSessionFactory" class="org.mybatis.spring.SqlSessionFactoryBean">
    <!-- 注入数据库连接池 -->
    <property name="dataSource" ref="dataSource" />
    <!-- 配置 MyBaties 全局配置文件 -->
    <property name="configLocation" value="classpath:mybatis-config.xml" />
    <!-- 扫描 entity 包,使用别名 -->
    <property name="typeAliasesPackage" value="cn.lrw.newsssm.entity" />
    <!-- 自动扫描 sql 配置文件 mapper 需要的 xml 文件 -->
    <property name="mapperLocations" value="classpath:mapper/*.xml" />
</bean>
```

（3）扫描 DAO 层接口，动态实现 DAO 接口、注入到 Spring 容器中。

```
<bean class="org.mybatis.spring.mapper.MapperScannerConfigurer">
    <!-- 注入 sqlSessionFactory -->
    <property name="sqlSessionFactoryBeanName" value="sqlSessionFactory" />
    <!-- 给出需要扫描 DAO 接口包 -->
    <property name="basePackage" value="cn.lrw.newsssm.dao" />
</bean>
```

5.5.3 配置 Spring-service

在 Service 层主要配置：

（1）自动扫描 cn.lrw.newsssm.service 包及其子包下面的 Java 文件，如果扫描到文件中带有@Service、@Component、@Repository、@Controller 等注解的类，则把这些类注册为 Bean。

```
<context:component-scan base-package="cn.lrw.newsssm.service"/>
```

（2）配置事务管理器，把事务管理交由 Spring 来完成。

```
<bean id="txManager" class="org.springframework.jdbc.datasource.DataSourceTransactionManager">
    <property name="dataSource" ref="dataSource" />
</bean>
```

（3）配置基于注解的声明式事务，可以直接在方法上加注解@Transaction。

```
<tx:annotation-driven transaction-manager="txManager" />
```

5.5.4 配置 Spring-web

在 Web 层主要配置：

（1）自动扫描 cn.lrw.newsssm.web 包及其子包下面的 Java 文件，如果扫描到文件中带有@Service、@Component、@Repository、@Controller 等注解的类，则把这些类注册为 Bean。

```
<context:component-scan base-package="cn.lrw.newsssm.web"/>
```

（2）开启 SpringMVC 注解模式，启用@RequestMapping、@ResponseBody 等注解，并将其注册到请求映射表中，处理请求的适配器确定调用哪个类的哪个方法，构造方法参数并返回值。

```
<mvc:annotation-driven />
```

（3）启用静态资源默认 Servlet 配置，与 web.xml 中配置<servlet-name>default</servlet-name>配合使用，加入对 js、css、jpg、gif、png 等静态资源的处理，允许使用“/”做整体映射。

```
<mvc:default-servlet-handler/>
```

（4）配置 JSP 视图解析器，例如在 controller 中某个方法返回一个 String 类型的“/admin”，实际上会返回“/WEB-INF/web/admin.jsp”。

```
<bean class="org.springframework.web.servlet.view.InternalResourceViewResolver" id=
"internalResourceViewResolver">
    <!-- 定义视图存放路径 -->
    <property name="prefix" value="/WEB-INF/web/" />
    <!-- 定义视图后缀 -->
    <property name="suffix" value=".jsp" />
</bean>
```

5.6 创建 DAO 接口

在 cn.lrw.newsssm.dao 包中创建 BaseDao 接口，主要实现数据库的操作。实际项目中，通常为每一个实体类创建一个独立的接口文件。当前项目简单，只创建了一个 BaseDao<T>，仍然按照增删改查的顺序，封装了一些常用的方法。**public interface** BaseDao<T>使用了泛型，减少了重复代码的编码。

1. 增

保存一个对象。

```
public int add1(T o) ;
public int add2(T o) ;
```

2. 删

删除一个对象。

```
public void delete(T o) ;
```

3. 改

更新一个对象。

```
public void update(T o) ;
```

4. 查

查询一个对象。

```
public T get1(@Param("id") int id) ;
public T get2(@Param("s1") String s1,@Param("s2") String s2) ;
```

查询一个集合。

```
public List<T> list1() ;
public List<T> list2(@Param("id")int pid) ;
```

分页查询一个集合，参数 s1 表示查询关键词、offset 表示第一行的偏移量、rows 表示每页的记录数。

```
public List<T> list3(@Param("offset") int page, @Param("rows") int rows) ;
public List<T> list4(@Param("s1") String s1,@Param("offset") int offset, @Param("rows") int rows) ;
```

查询记录数。

```
public Long countofUser() ;
public Long countofNews() ;
```

5.7　创建 Mapper 文件

实际项目中，需要在 resources\mapper 文件夹中，为每一个接口创建一个独立的 Mapper 映射文件*.xml，每一个接口中的方法都必须在 mapper.xml 中的<mapper>标签中对应着有一个 id 的唯一的 Bean，在 Bean 标签里面写对应的 SQL 语句。

映射文件*.xml 由 Mapper 标签开始，由/mapper 结束。属性 namespace（命名空间）主要在代理中使用，namespace 是唯一的，是该 xml 对应的接口全名，它把 Mapper 标签和接口联系在一起。Mapper 的目的是为 DAO 接口方法提供 SQL 语句配置。

标签 select、insert、updatet 和 delete 的 id 属性对应 DAO 接口中的方法名；属性 resultType 是返回值类型；属性 parameterType 是参数类型；SQL 语句中的占位符#{...}中填写的是方法的参数。

如果返回值是自定义实体类的对象，则让 select 元素使用属性 resultMap。假定有一个自定义类 EStudent，如下定义：

```
public class EStudent {
    private long id;
    private String name;
    private int age;
    // getter，setter 方法
    //必须提供一个无参数的构造函数
    public EStudent(){}
}
```

新建一个 resultMap 节点，建立 SQL 查询结果字段与实体属性的映射关系信息。resultMap 的属性 id 唯一标识 resultMap；属性 type 映射实体类的类名或别名；子元素 id 用于设置主键字段与模型属性的映射关系；子元素 result 用于设置普通字段与模型属性的映射关系；子元素的 column 属性表示库表的字段名；子元素的 property 属性表示实体类里的属性名。以下配置信息只是 resultMap 的一个应用样例，不在当前项目中应用。

```
<resultMap id="getStudentRM" type="EStudnet">
  <id property="id" column="ID"/>
  <result property="studentName" column="Name"/>
  <result property="studentAge" column="Age"/>
</resultMap>
<select id="getStudent" resultMap="getStudentRM">
  SELECT ID, Name, Age
    FROM TStudent
</select>
```

映射文件 mMapper.xml 的内容如下，分别针对 User、News、CMenu 类，每个类按照增删改查的顺序配置 SQL。

```
<?xml version="1.0" encoding="UTF-8"?>
<!DOCTYPE mapper PUBLIC "-//mybatis.org//DTD Mapper 3.0//EN"
```

```
"http://mybatis.org/dtd/mybatis-3-mapper.dtd" >
<mapper namespace="cn.lrw.newsssm.dao.BaseDao">
<!-- =============用户操作：增 ==================== -->
    <insert id="add1" parameterType="User" >
        insert into user (uid,xm,pwd,bj,role ) values (#{uid},#{xm},#{pwd},#{bj},#{role})
    </insert>
<!-- =============用户操作：查==================== -->
    <select id="get2" resultType="User">
        select * from user
        <where> uid=#{s1} and pwd=#{s2} </where>
    </select>
    <select id="countofUser" resultType="Long">
        select count(*) from user
    </select>

<!-- =============新闻操作：增 ==================== -->
    <insert id="add2" parameterType="News" >
        insert into news ( title,content,tjdate,cruser,hitnum ) values (#{title},#{content},#{tjdate},
#{cruser}, #{hitnum})
    </insert>
<!-- =============新闻操作：删 ==================== -->
    <delete id="delete" >
        delete from news where id= #{id}
    </delete>
<!-- =============新闻操作：改 ==================== -->
    <update id="update" parameterType="News">
        update news set title=#{title},content=#{content},cruser=#{cruser},hitnum=#{hitnum} where
id=#{id}
    </update>
<!-- =============新闻操作：查 ==================== -->
    <select id="get1" resultType="News">
        select * from news   where id= #{id}
    </select>
    <select id="list3" resultType="News">
        select * from news order by id desc limit ${offset},${rows}
    </select>
    <select id="list4" resultType="News">
        select * from news
        <where>
            title like   '%${s1}%'
        </where>
         order by id desc limit ${offset},${rows}
    </select>
    <select id="countofNews" resultType="Long">
        select count(*) from news
    </select>
```

```
<!-- =============菜单操作：查 ======================= -->
    <select id="list2" resultType="Cmenu">
        select * from cmenu where pid=${id}
    </select>
</mapper>
```

5.8　公共方法类

在业务逻辑类和控制器类中可能会多次用到一些相同的方法，开发人员通常会把这样的方法封装成工具类 BaseUtil，存放于 cn.lrw.newsssm.utils 包中，类中方法的定义请参照 4.7 节公共方法类。

5.9　创建业务逻辑类

业务逻辑类是控制器类与 DAO 之间的桥梁，类中方法被控制器调用，方法内部进行业务处理，调用 DAO 方法访问数据库，类中用到几个注解：

- @Service，标注业务层组件。
- @Resource，把 DAO 注入到 service 中，不需要 new 一个对象。

5.9.1　UserSvc 类

在 cn.lrw.newsssm.service 包中创建 UserSvc 类，实现添加新用户、查询用户或用户数量等业务逻辑。

```
@Service
public class UserSvc {
    @Resource
    private BaseDao<User> dao;
    public void addU(User user) {
        dao.add1(user);
    }
    public User findU(String uid, String pwd) {
        return dao.get2(uid,pwd);
    }
    public Long getCount() {
        return dao.countofUser();
    }
}
```

5.9.2　NewsSvc 类

在 cn.lrw.newsssm.service 包中创建 NewsSvc 类，实现新闻信息增删改查的业务逻辑。

```
@Service
public class NewsSvc {
    @Resource
```

```
    private BaseDao<News> dao;
    //===========增================
    public void addNews(News news) throws Exception{
        dao.add2(news);
    }
    //===========删================
    public void deleteNews(int id) throws Exception{
        News u =(News) dao.get1(id);
        dao.delete(u);
    }
    //===========改================
    public void updateNews(News news) throws Exception{
        dao.update(news);
    }
    //===========查================
    //按新闻标题分页查询
    public List<News> listDgNews(String title,int page,int rows){
        if (page < 1) {
            page = 1;
        }
        if (rows < 1) {
            rows = 10;
        }
        page=(page-1)*rows;
        if(title == null || "".equals(title)) return dao.list3(page, rows);
        else return dao.list4(title, page, rows);
    }
    //根据新闻 id 查询新闻，用于阅读新闻
    public News getNews( int id){
        News news=dao.get1( id);
        news.setHitnum(news.getHitnum()+1);   //单击量增加
        dao.update(news);
        return   news;
    }
    //查询新闻数量
    public int getNewsCount(){
        try{
            Long a=dao.countofNews();
            return   Integer.parseInt(a.toString());
        }catch(Exception e){
            e.printStackTrace();
            return 0;
        }
    }
}
```

5.9.3　MenuSvc 类

在 cn.lrw.newsssm.service 包中创建 MenuSvc 类，实现 Tree 形菜单数据的获取。

```
@Service
public class MenuSvc {
    @Resource
    private BaseDao<Cmenu> dao;
    public List<Cmenu> listMenu(int pid){
        return dao.list2(pid);
    }
}
```

5.10　创建控制器类

控制器类负责接收前端的请求，根据情况进行转发，将处理结果回送前端。用到的几个注解：

- @Controller，标注当前类为控制器类。
- @RequestMapping，是 Spring Web 应用程序中最常被用到的注解之一，它将 HTTP 请求映射到 MVC 和 REST 控制器或控制器的处理方法上。
- @Autowired，对类成员变量、方法及构造函数进行标注，完成自动装配的工作，通过 @Autowired 的使用来消除 setter/getter 方法。

5.10.1　UserAct 类

在 cn.lrw.newsssm.web 包中创建 UserAct 类，实现用户登录、退出和跳转到后台的请求。

```
@Controller
@RequestMapping("/user")
public class UserAct{
    @Autowired
    private UserSvc userSvc;
    private HashMap<String,Object> jsonobj=new HashMap<String,Object>();
    @RequestMapping(value="/doLogin",method=RequestMethod.POST)
    public void doLogin(String uid,String pwd,HttpSession session,HttpServletResponse response,
Model model){
        try {
            Long n=userSvc.getCount();
            if(n==0){
                User user=new User();
                user.setUid("2953");
                user.setXm("lrw");
                user.setPwd(BaseUtil.lrwCode("123", ""));
                user.setRole("1");
                userSvc.addU(user);
            }
            pwd=BaseUtil.lrwCode(pwd, "");
```

```
                User user0 = userSvc.findU(uid, pwd);
                jsonobj.clear();
                if(user0 != null){
                    jsonobj.put("ok", true);
                    jsonobj.put("msg", "user/goIndex");
                    session.setAttribute("me", user0);
                }else
                    {
                    jsonobj.put("ok", false);
                    jsonobj.put("msg", "用户不存在");
                    }
            } catch (Exception e) {
                jsonobj.put("ok", false);
                jsonobj.put("msg", "系统错误 2");
            }
            BaseUtil.outPrint(response, BaseUtil.toJson(jsonobj));
        }
        @RequestMapping(value="/doLogout",method=RequestMethod.GET)
        public String doLogout(HttpSession session){
            session.invalidate();
            return "redirect:/index.jsp";
        }
        @RequestMapping(value="/goIndex")
        public String goIndex(){
            return "/admin";
        }
    }
```

5.10.2 NewsAct 类

在 cn.lrw.newsssm.web 包中创建 NewsAct 类，实现对新闻进行增删改查的请求。

```
@Controller
@RequestMapping("/news")
public class NewsAct   {
    @Autowired
    private NewsSvc newsSvc;
    private String jsonResult;
    private HashMap<String,Object> jsonobj=new HashMap<String,Object>();
//===========增================
    @RequestMapping(value="/goAdd")
    public String goAdd(){
        return "/newsadd";
    }
    @RequestMapping(value="/saveAdd",method=RequestMethod.POST)
    public void saveAdd(News news,HttpServletResponse response){
        jsonobj.clear();
        try {
```

```
            news.setTjdate(new Date());  //提交日期由后端生成
            news.setHitnum(0);
            newsSvc.addNews(news);
            jsonobj.put("ok", true);
            jsonobj.put("msg", "goadmin");
        } catch (Exception e) {
            e.printStackTrace();
            jsonobj.put("ok", false);
            jsonobj.put("msg", "系统错误 2");
        }
        jsonResult = BaseUtil.toJson(jsonobj);
        BaseUtil.outPrint(response, jsonResult);
    }
//===========删===============
    @RequestMapping(value="/doDel1",method=RequestMethod.POST)
    public void doDel1(int id,HttpServletResponse response){
        jsonobj.clear();
        boolean delflag=false;
        try{
            newsSvc.deleteNews(id);
            delflag=true;
        }catch(Exception e){
            e.printStackTrace();
            delflag=false;
        }
        jsonobj.put("delflag", delflag);
        BaseUtil.outPrint(response, BaseUtil.toJson(jsonobj));
    }
//===========改===============
    @RequestMapping(value="/goEdit",method=RequestMethod.GET)
    public String goEdit(int id,Model model){
        model.addAttribute("news", newsSvc.getNews(id));
        return "/newsedit";
    }
    @RequestMapping(value="/saveEdit",method=RequestMethod.POST)
    public void saveEdit(News news,HttpServletResponse response){
        jsonobj.clear();
        try {
            News news0=newsSvc.getNews( news.getId());
            news0.setContent(news.getContent());
            news0.setCruser(news.getCruser());
            news0.setTitle(news.getTitle());
            newsSvc.updateNews(news0);
            jsonobj.put("ok", true);
            jsonobj.put("msg", "goadmin");
        } catch (Exception e) {
```

```
                e.printStackTrace();
                jsonobj.put("ok", false);
                jsonobj.put("msg", "系统错误 2");
            }
            jsonResult = BaseUtil.toJson(jsonobj);
            BaseUtil.outPrint(response, jsonResult);
        }
//===========查===============
        @RequestMapping(value="/goList")
        public String goList(){
            return "/newslist";
        }
        @RequestMapping(value="/getCount",method=RequestMethod.POST)
        public void getCount(HttpServletResponse response){
            int c=0;
            try{
                c=newsSvc.getNewsCount();
            }catch(Exception e){
                e.printStackTrace();
                c=0;
            }
            jsonobj.clear();
            jsonobj.put("newscount", c);
            jsonResult = BaseUtil.toJson(jsonobj);
            BaseUtil.outPrint(response, jsonResult);
        }
        @RequestMapping(value="/getaNews",method=RequestMethod.GET)
        public String getaNews(int id,Model model){
            model.addAttribute("news",newsSvc.getNews(id));
            return "/newsread";
        }
        @RequestMapping(value="/listNews",method=RequestMethod.POST)
        public void listNews(String s_name,int page,int rows,HttpServletResponse response){
            List<News> newslist=newsSvc.listDgNews(s_name,page,rows);
            jsonobj.clear();
            jsonobj.put("total", newslist.size());
            jsonobj.put("rows", newslist);
            jsonResult = BaseUtil.toJson(jsonobj);
            BaseUtil.outPrint(response, jsonResult);
        }
    }
```

5.10.3 MenuAct 类

在 cn.lrw.newsssm.web 包中创建 MenuAct 类，实现对菜单数据的封装。

```
    @Controller
    public class MenuAct   {
```

```
    @Autowired
    private MenuSvc menuSvc;
    private String jsonResult;
    @RequestMapping(value="menutree")
    public void menutree(HttpServletRequest req,HttpServletResponse response) {
        HttpSession session=req.getSession();
        User user = (User) session.getAttribute("me");
        String role=user.getRole();
        //====================一级菜单==========================
        List<Cmenu> menulist=menuSvc.listMenu(0);
        List<EasyUITree> eList = new ArrayList<EasyUITree>();
        if(menulist.size() != 0){
            for (int i = 0; i < menulist.size(); i++) {
                Cmenu t = menulist.get(i);
                if(!t.getPermission().contains(role))continue;
                EasyUITree e = new EasyUITree();
                e.setId(t.getId()+"");
                e.setText(t.getName());
                List<EasyUITree> eList2 = new ArrayList<EasyUITree>();
            //====================二级菜单==========================
                List<Cmenu> menu2 = menuSvc.listMenu(t.getId());
                for (int j = 0; j < menu2.size(); j++) {   //二级菜单
                    Cmenu t2 = menu2.get(j);
                    if(!t2.getPermission().contains(role))continue;
                    Map<String,Object> attributes = new HashMap<String, Object>();
                    attributes.put("url", t2.getUrl());
                    attributes.put("role", t2.getPermission());
                    EasyUITree e1 = new EasyUITree();
                    e1.setAttributes(attributes);
                    e1.setId(t2.getId()+"");
                    e1.setText(t2.getName());
                    e1.setState("open");
                    eList2.add(e1);
                }
                e.setChildren(eList2);
                e.setState("closed");
                eList.add(e);
            }
        }
        jsonResult = BaseUtil.toJson(eList);
        BaseUtil.outPrint(response, jsonResult);
    }
}
```

5.11 文件上传类

在 cn.lrw.newsssm.utils 包中创建 FileAct 类，实现前端页面使用 UEditor 在线编辑器上传文件时后端接收并写入磁盘。

```
public class FileAct extends HttpServlet {
    public FileAct(){
        super();
    }
    protected void doGet(HttpServletRequest request, HttpServletResponse response) throws
    ServletException, IOException {
        doPost(request, response);
    }
    protected void doPost(HttpServletRequest request, HttpServletResponse response) throws
    ServletException, IOException {
        request.setCharacterEncoding( "UTF-8" );
        response.setHeader("Content-Type" , "text/html");
        ServletContext application=request.getServletContext();
        String rootPath = application.getRealPath( "/" );
        PrintWriter out = response.getWriter();
        out.write( new ActionEnter( request, rootPath ).exec() );
    }
}
```

5.12 前端页面

由于当前项目中的前端页面均使用 JSP，如果通过向导创建了 JSP 页面文件，则在文件开头默认定义了全局变量 basePath，表示当前项目的 URL，如 http://localhost:8080/newsssm/。参照 4.13 节完成前端页面。以下的每一小节，主要列出有变动的内容，不再重复全部编码。实际上，变化最大的是向后端请求的 URL。

5.12.1 系统首页

index.jsp 是系统首页，分页显示新闻列表和登录窗口。首页位于 WebRoot 目录下，可以直接访问，所以可以使用相对路径引用 JS、CSS 文件和图片文件。详细的实现方法，请参照 3.5.1 美化系统首页、3.5.2Ajax 方法、3.5.3 更友好的 alert、3.5.4 标题图标、3.8.10.1 前台新闻列表。

页面代码已在 4.13.1 节提供，但需要修改 login.js 中向后端请求的 URL，如：

- 新闻数量的获取。

```
url:"./news/getCount"
```

- 新闻列表的获取。

```
url:"./news/listNews"
```

- 动态加载每一条新闻时的超链接。

```
<a href='./news/getaNews
```

- 请求登录。
  url : base+"user/doLogin"

5.12.2　出错跳转页

当出现系统错误时，直接跳转到 nologin.jsp 页，页面内容已在 4.13.2 节提供。

5.12.3　新闻阅读页

在新闻阅读页 newsread.jsp，尝试使用 jstl 数据标签，而在 4.13.3 节中列出的新闻阅读页使用了 Struts 的标签。

```
<%@ page language="java" import="java.util.*" pageEncoding="UTF-8"%>
<%
String path = request.getContextPath();
String basePath = request.getScheme()+"://"+request.getServerName()+":"+request.getServerPort()+
path+"/";
%>
<%@ taglib uri="http://java.sun.com/jsp/jstl/fmt" prefix="fmt"%>
<!DOCTYPE html>
<html>
<head>
    <meta charset="UTF-8" />
    <title>读新闻</title>
    <link rel="shortcut icon" href="<%=basePath%>include/img/logo.png" type="image/x-icon" />
    <link rel="stylesheet" type="text/css" href="<%=basePath%>include/css/news.css">
    <script type="text/javascript" src="<%=basePath%>include/js/jquery.min.js"></script>
</head>
<body>
    <div style="background:#B3DFDA;padding:0 10px 0 10px;vertical-align: middle;">
        <img src="<%=basePath%>include/img/logo.png" width="126" height="50" />
        <div style="float:right;line-height:50px;margin-right:10px;font-size: 9pt;">
            <span>【</span><a style="color:blue;" href="javascript:window.close();"><span>
            关闭窗口</span></a><span>】</span>
        </div>
    </div>
<div class="ndetail">
    <div class="ntitle">${news.title}</div>
    <div class="nauthor">
        <div>来源:  <strong>${news.cruser}</strong>  发布时间: 
            <strong><fmt:formatDate value="${news.tjdate}" pattern="yyyy-MM-dd"/></strong>
              访问量:  <strong>[<span>${news.hitnum}</span>]</strong></div>
    </div>
    <div class="nbody">
        <div id="vsb_content"> ${news.content}</div>
    </div>
</div>
</body>
</html>
```

5.12.4 后台 Layout

新闻管理的后台布局页 admin.jsp，页面代码参照 4.13.4 节，主要修改了 JS 部分对后端请求的 URL。

- 对树形菜单数据的请求。

```
url : "<%=basePath%>menutree",
```

- 退出系统。

```
top.location.href="<%=basePath%>user/doLogout";
```

- 默认新闻列表页。

```
swNewTab('新闻管理',"<%=basePath%>news/goList");
```

5.12.5 新闻列表页

新闻列表页 newslist.jsp 页面上使用 datagrid 加载分页新闻。每条新闻后有修改和删除的链接，工具栏中有标题关键字输入框、查询按钮。页面内容参照 4.13.5 节，主要修改了 JS 部分对后端请求的 URL。

- 新闻列表的获取。

```
url:"<%=basePath%>news/listNews",
```

- 请求修改新闻。

```
parent.swNewTab("修改新闻信息","<%=basePath%>news/goEdit?id="+id);
```

- 请求删除新闻。

```
url:"<%=basePath%>news/doDel1",
```

5.12.6 新闻添加页

新闻添加页 newsadd.jsp 页面，使用百度在线编辑器 UEditor，页面内容参照 4.13.6 节。

- 修改 ueditor.config.js 配置中文件上传服务器 URL。

```
serverUrl: base+"bdupfile/upload",
```

- 把百度编辑器的配置文件 config.json，放在当前项目的 resources 目录下。
- 修改 newsadd.jsp 页面中发布新闻的请求及数据封装方式的 JS 代码。

```
url : "<%=basePath%>news/saveAdd",
data : {"title":a,"content":b,"cruser":c},
```

5.12.7 新闻修改页

新闻修改页 newsedit.jsp，页面内容参照 4.13.7 节，仅仅修改了 JS 代码中保存修改后新闻的 URL。

```
url : "<%=basePath%>news/saveEdit",
```

5.13 增强安全

可以在 cn.lrw.newsssm.utils 包中设置一个自定义的过滤器 LoginFilter 来拦截未登录用户的某些请求，定义内容参照 4.14.1 节。

5.14　配置 web.xml

在 web.xml 中，除了已经配置的 Spring 监听、SpringMVC、UTF-8 编码过滤器以外，还需要配置以下的内容。

- 实现百度上传文件功能的 servlet。

```
<!-- 允许百度编辑器 UEditor 的上传请求 -->
<servlet>
    <servlet-name>upload</servlet-name>
    <servlet-class>cn.lrw.newsssm.utils.FileAct</servlet-class>
</servlet>
<servlet-mapping>
    <servlet-name>upload</servlet-name>
    <url-pattern>/bdupfile/upload</url-pattern>
</servlet-mapping>
```

- 开启默认的 servlet（对静态资源）。

```
<!-- 允许用户对静态资源的请求 -->
<servlet-mapping>
    <servlet-name>default</servlet-name>
    <url-pattern>*.js</url-pattern>
    <url-pattern>*.css</url-pattern>
    <url-pattern>/error/*"</url-pattern>
    <url-pattern>/include/*</url-pattern>
    <url-pattern>/upload/*</url-pattern>
</servlet-mapping>
```

- 增强安全的 loginFilter 过滤器（包含不需要过滤 excludeStrings 的字符串）。

```
<!-- 拦截验证用户的登录信息 -->
<filter>
    <filter-name>loginFilter</filter-name>
    <filter-class>cn.lrw.newsssm.utils.LoginFilter</filter-class>
    <init-param>
        <param-name>excludeStrings</param-name><!-- 对登录页面不进行过滤 -->
<param-value>doLogin;listNews;getaNews;getCount;/include;/error;/upload;/index.jsp</param-value>
    </init-param>
    <init-param>
        <param-name>redirectPath</param-name><!-- 未通过,跳转到报错界面 -->
        <param-value>/error/nologin.jsp</param-value>
    </init-param>
</filter>
<filter-mapping>
    <filter-name>loginFilter</filter-name>
    <url-pattern>/*</url-pattern>
</filter-mapping>
```

5.15 考核任务

正确实现基于 SSM 框架的新闻信息增删改查，添加与修改新闻时可以上传图片等文件。

（1）添加新闻。（20 分）

（2）修改新闻。（20 分）

（3）后台新闻列表。（20 分）

（4）删除新闻。（20 分）

（5）前台新闻列表。（10 分）

（6）阅读新闻。（10 分）

本章小结

本章开发基于 SSM（Spring+SpringMVC+MyBatis）框架的 JavaEE 项目，其中 Spring 是一个轻量级的控制反转（IoC）和面向切面（AOP）的容器框架。SpringMVC 分离了控制器、模型对象、分派器以及处理程序对象的角色，这种松耦合让每一层都更容易进行定制。MyBatis 是一个支持普通 SQL 查询、存储过程和高级映射的优秀持久层框架。

系统运行的基本流程是，页面发送请求给控制器，控制器调用业务层处理逻辑，逻辑层向持久层发送请求，持久层与数据库交互后将结果返回给业务层，业务层将处理结果发送给控制器，控制器再调用视图展现数据。

基于 SSM 框架项目的工作原理如图 5-9 所示，以请求登录为例。

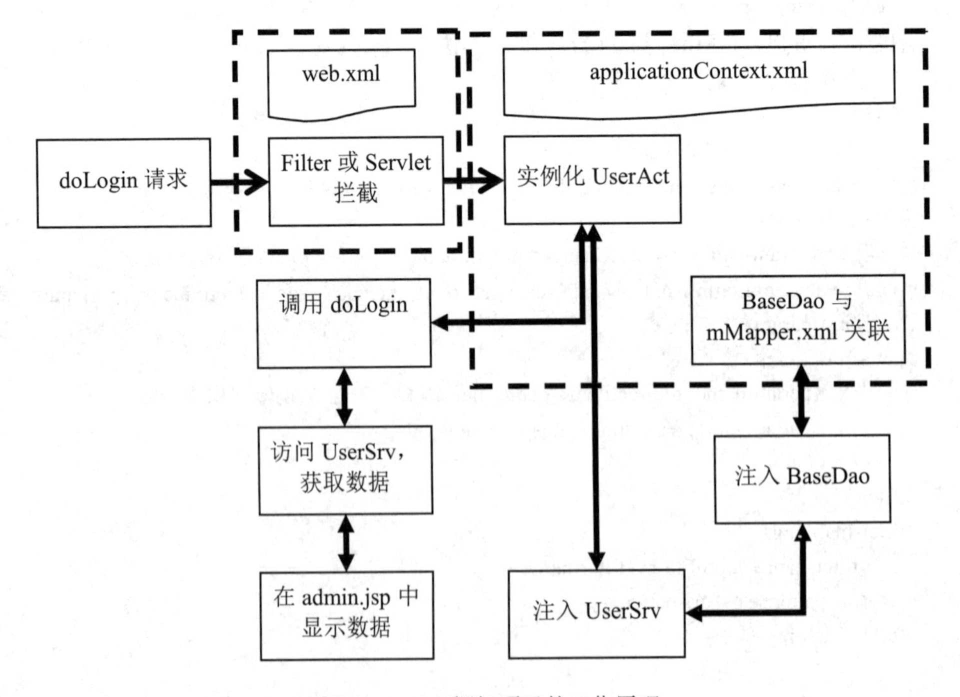

图 5-9 SSM 框架项目的工作原理

（1）在浏览器上访问路径 /user/doLogin。

（2）Tomcat 根据 web.xml 上的配置信息拦截到了/doLogin，并将其交由 DispatcherServlet 处理。

（3）DispatcherServlet 根据 SpringMVC 的配置，将请求交由 UserAct 类的实例进行处理。

（4）在实例化 UserAct 的时候，注入 UserService。

（5）在实例化 UserService 的时候，又注入 BaseDao。

（6）根据 applicationContext.xml 中的配置信息，将 BaseDao 和 mMapper.xml 关联起来。

（7）有了实例化的 UserAct，并调用 doLogin 方法。

（8）在 doLogin 方法中，访问 UserService 并获取数据，回送到前端页面。前端页面再请求跳转到后台管理页面 admin.jsp，接着服务端响应实现跳转到 admin.jsp。

（9）最后在浏览器显示 admin.jsp 页面。

第 6 章　基于 JFinal 的项目实战

本章介绍采用前端与后端分离开发的方式来开发基于 JFinal 的项目，首先搭建 JFinal 框架的 Web 项目；然后编写后端的 Java 类：model、service、controller、interceptor，配置系统 SysConfig，创建 Mapper 映射文件；最后参照第 3 章的前端页面开发方法快速开发当前项目的前端页面。

6.1　JFinal 简介

JFinal 是基于 Java 语言的极速 WEB + ORM 框架。其核心设计目标是开发迅速、代码量少、学习简单、功能强大、轻量级、易扩展。

JFinal 有如下主要特点：

（1）MVC 架构，设计精巧，使用简单。

（2）遵循全称原则，零配置，无 xml。

（3）独创 Db + Record 模式，灵活便利。

（4）ActiveRecord 支持，使数据库开发极致快速。

（5）自动加载修改后的 Java 文件，开发过程中无需重启 Web Server。

（6）AOP 支持，拦截器配置灵活，功能强大。

（7）Plugin 体系结构，扩展性强。

（8）多视图支持，支持 FreeMarker、JSP、Velocity。

（9）强大的 Validator 后端校验功能。

（10）功能齐全，拥有 Struts2 的绝大部分功能。

（11）体积小，仅 632K，且无第三方依赖。

JFinal 采用微内核全方位扩展架构，全方位是指其扩展方式在空间上的表现形式。JFinal 由 Handler、Interceptor、Controller、Render、Plugin 五大部分组成。JFinal 顶层架构如图 6-1 所示。

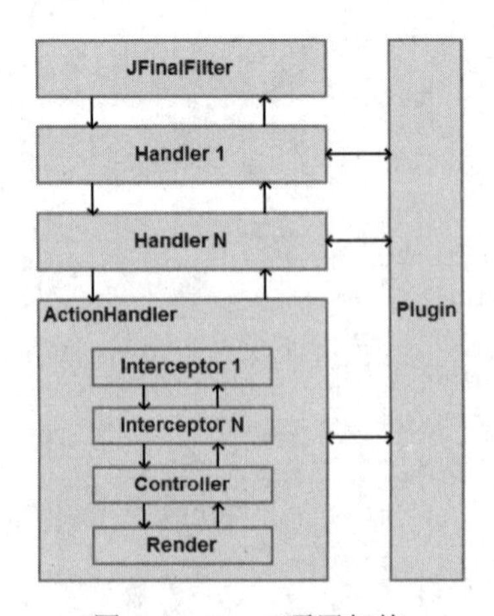

图 6-1　JFinal 顶层架构

6.2　创建 JFinal 项目

创建项目前，先设计好项目工程结构、准备好所需的 Jar 包、JS 库、CSS 库、MySQL 数据库等资源。

6.2.1　项目工程结构

表 6-1　项目工程结构

newsjfinal	项目名称
├─src	源码文件夹
│ ├─cn.lrw.newsjfinal	主包，存放项目配置类
│ ├─cn.lrw.newsjfinal.model │ ├─cn.lrw.newsjfinal.model.base	存放数据库表/视图对应的 Java 类
│ ├─cn.lrw.newsjfinal.controller	存放控制器类
│ ├─cn.lrw.newsjfinal.service	存放业务逻辑类
│ ├─cn.lrw.newsjfinal.interceptor	存放拦截器类
│ └─cn.lrw.newsjfinal.utils	存放封装公用方法或属性的类
├─res	存放 JSON、Properties 等类型的配置文件
│ ├─log4j.properties	系统日志配置文件
│ ├─config.json	UEditor 的配置文件
│ └─beetl.properties	Beetl 模板配置文件
└─WebRoot	Web 项目根目录
├─error	存放异常访问提示页，如 nologin.html、403.html、404.html 等
├─include	分类存放网页中引用的 JS、CSS、图像类文件
│ ├─css	
│ ├─js	
│ ├─img	
│ ├─easyui	EasyUI 框架
│ └─ueditor	可视化 HTML 编辑器
├─upload	存放上传的文件
├─WEB-INF	Java Web 应用的安全目录，自动存放 class 文件，客户端无法访问
│ ├─lib	存放项目需要的 Jar 包
│ ├─html	存放前端网页文件，如 HTML、JSP 等
│ └─web.xml	Web 工程的配置文件
└─index. html	Web 工程默认的首页文件

6.2.2 准备 Jar 包和 JS 库

本项目使用表 6-2 所示的的 Jar 包，如果需要其他的 Jar 包，可以从 Maven 查找和下载。

表 6-2 项目所需 Jar 包

Jar 包	说明
JFinal	基于 Java 语言的极速 WEB + ORM 框架
mysql-connector-java	MySQL 数据库驱动，较低版本，如 5.1.39
druid	数据库连接池 Druid，带强大的监控功能
Log4j	用 Java 编写的可靠，快速和灵活的日志框架（API）
beetl	一款 6 倍于 Freemarker 的超高性能的 Java 模板引擎
shiro	用于认证、授权、加密、会话管理、与 Web 集成、缓存等
ueditor	百度在线编辑器。依赖 json.jar、commons-codec-1.10.jar、commons-io-2.4.jar、commons-fileupload-1.3.3.jar、commons-lang3-3.2.1.jar

本项目还将使用到一些常用的 JS 库文件，如 jQuery，EasyUI，UEditor 等。

6.2.3 新建 web 项目

新建 Web Project，输入适当的项目名称，比如 newsjfinal，选择适当的环境参数如图 6-2 所示。由于使用 Beetl 模板，不需要 JSTL。

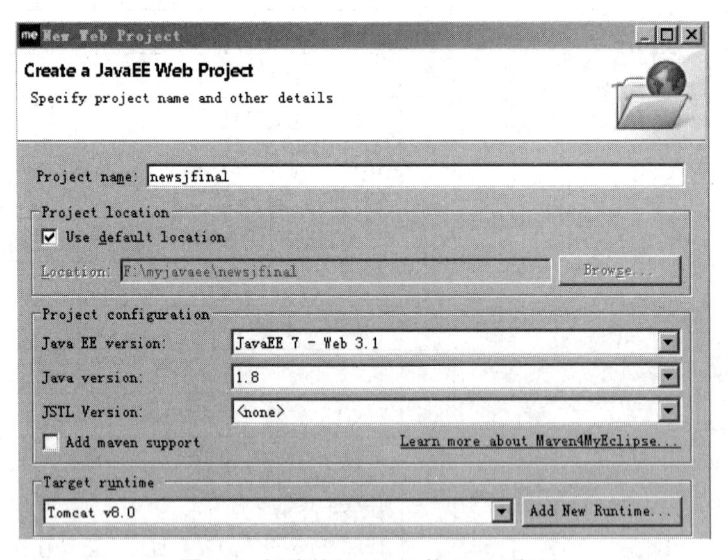

图 6-2 新建基于 JFinal 的 Web 项目

添加源文件夹 res，用于存放配置文件，如图 6-3 所示。

勾选生成 web.xml，不需要 index.jsp，所以不要勾选生成 index.jsp。

创建完成之后，把准备好的 Jar 包添加到 WebRoot/WEB-INF/lib 目录下，如图 6-4 所示。

参照“项目工程结构”添加包（package）和文件夹，如图 6-5 所示。

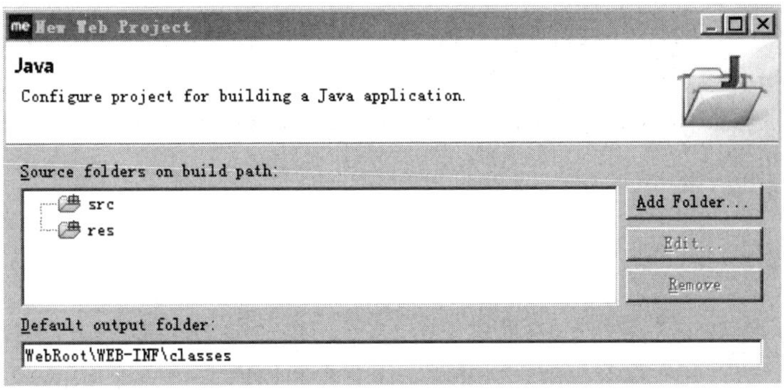

New Web Project

Java

Configure project for building a Java application.

Source folders on build path:

src

res

Add Folder...

Edit...

Remove

Default output folder:

WebRoot\WEB-INF\classes

图 6-3 添加 res 源文件夹

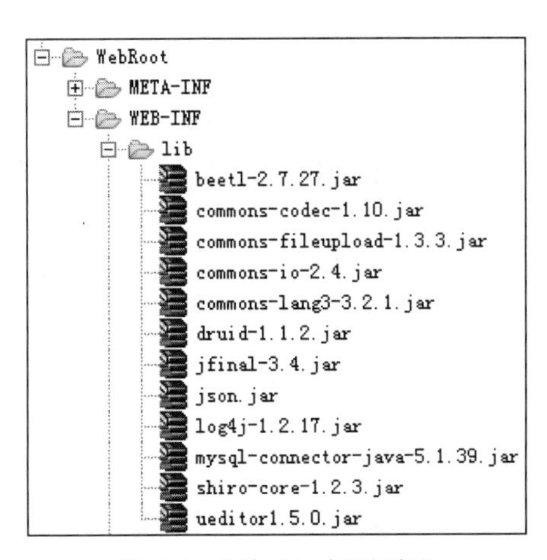

图 6-4 添加 Jar 包到项目

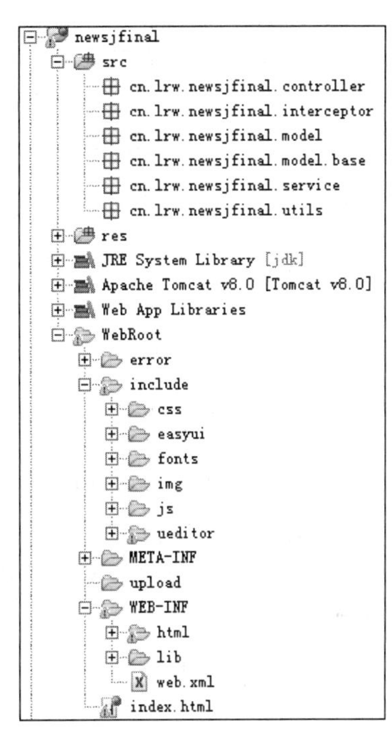

图 6-5 项目工程结构

6.2.4 添加数据源

利用可视化工具 HeidiSQL，创建数据库 dbnews4。在 dbnews4 数据库中，根据 2.3 节所示的表结构创建数据库表 user、news 和 cmenu，手工输入表 cmenu 的数据，如图 6-6 所示。

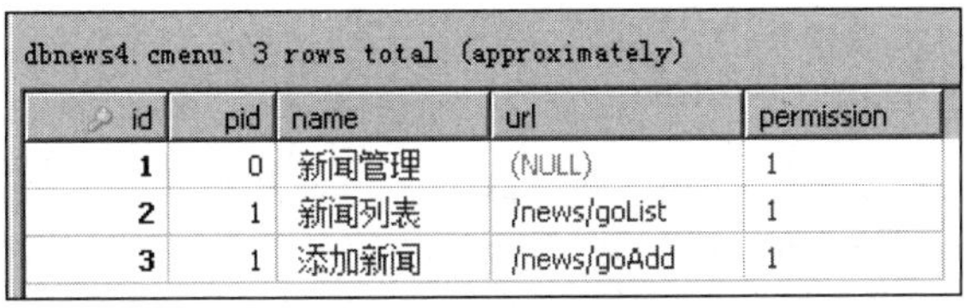

dbnews4.cmenu: 3 rows total (approximately)

id	pid	name	url	permission
1	0	新闻管理	(NULL)	1
2	1	新闻列表	/news/goList	1
3	1	添加新闻	/news/goAdd	1

图 6-6 表 cmenu 数据

在工程结构的 res 文件夹中创建 db.properties 文件，配置数据库的连接，如果需要密码连接数据库，则在“password =”后加上密码。

```
url = jdbc:mysql://127.0.0.1:3066/dbnews4?serverTimezone=Hongkong&useSSL=false
user = root
password =
```

6.2.5 组件 Model

ActiveRecord 是 JFinal 最核心的组成部分之一，通过 ActiveRecord 来操作数据库，将极大地减少代码量，提升开发效率。ActiveRecord 模式的核心是：一个 Model 对象唯一地对应数据库表中的一条记录。而对应关系依靠的是数据库表的主键值，所以要求数据库表必须要有主键。

Model 是 ActiveRecord 中最重要的组件之一，它充当 MVC 模式中的 Model 部分。假如有一个 Bean 类 User，User 继承 Model，则立即拥有许多操作数据库的方法。在 User 中，可以声明一个静态对象 dao，方便查询操作，但该对象并不是必须的。基于 ActiveRecord 的 Model 无需定义属性，无需定义 getter 和 setter 方法，无需 XML 配置，无需 Annotation 配置，极大减少了代码量。

Model 与 Bean 相结合，具有很多的优势：

- 充分利用海量的针对于 Bean 设计的第三方工具，例如 jackson。
- 快速响应数据库表变动，极速重构，提升开发效率，提升代码质量。
- 拥有 IDE 代码提示，不用记忆数据表字段名，消除记忆负担，避免手写字段名出现手误。
- BaseModel 设计令 Model 中依然保持清爽，在表结构变化时极速重构关联代码。
- 自动化 Table 至 Model 映射。
- 自动化主键、复合主键名称识别与映射。
- MappingKit 承载映射代码，JFinalConfig 保持干净清爽。
- 有利于分布式场景和无数据源时使用 Model。
- 新设计避免了以往自动扫描映射设计的若干缺点：引入新概念（如注解）增加学习成本、性能低、Jar 包扫描可靠性与安全性低。

以下是 Model 中操作数据库的一些常见用法，按照增删改查的顺序，分别以样例说明。

1. 增

创建 name 属性为 James，age 属性为 25 的 User 对象并添加到数据库。

```
new User().set("name", "James").set("age", 25).save();
```

2. 删

删除 id 值为 25 的 User。

```
User.dao.deleteById(25);
```

3. 改

查询 id 值为 25 的 User，将其 name 属性改为 James 并更新到数据库。

```
User.dao.findById(25).set("name", "James").update();
```

4. 查

查询 id 值为 25 的 user，仅仅取其 name 与 age 两个字段的值。

```
User user = User.dao.findByIdLoadColumns(25, "name, age");
```

获取 user 的 name 属性。

```
String userName = user.getStr("name");
```

获取 user 的 age 属性。

```
Integer userAge = user.getInt("age");
```

查询所有年龄大于 18 岁的 user。

```
List<User> users = User.dao.find("select * from user where age>18");
```

分页查询年龄大于 18 的 user，当前页号为 1，每页 10 个 user。

```
Page<User> userPage = User.dao.paginate(1, 10, "select *", "from user where age > ?", 18);
```

SQL 最外层带 group by 的 paginate。

```
dao.paginate(1, 10, true, "select *", "from girl where age > ? group by age", 18);
```

将查询总行数与查询数据的两条 SQL 独立出来，这样处理主要是应对具有复杂 order by 语句或者 select 中带有 distinct 的情况。

```
String from = "from girl where age > ?";
String totalRowSql = "select count(*) " + from;
String findSql = "select * " + from + " order by age";
dao.paginateByFullSql(1, 10, totalRowSql, findSql, 18);
```

6.2.6　生成器 Generator

ActiveRecord 模块的 com.jfinal.plugin.activerecord.generator 包下，提供了一个 Generator 工具类，可自动生成 Model、BaseModel、MappingKit、DataDictionary 四类文件。利用生成器 Generator 生成的 Model 与 JavaBean 合体，立即拥有了 getter 和 setter 方法，遵守传统的 JavaBean 规范。

创建生成器的步骤：

（1）在 cn.lrw.newsjfinal.model 包中创建带有 main 函数的 GeneratorModel 类。

（2）在 main 函数中，定义四个参数：

- baseModelPackageName 表示 baseModel 的包名。
- baseModelOutputDir 表示 baseModel 的输出路径。
- modelPackageName 表示 Model 的包名（MappingKit 默认使用的包名）。
- modelOutputDir 表示 Model 的输出路径（MappingKit 文件默认的保存路径）。

（3）根据 db.properties 生成 DruidPlugin 对象 cp。

（4）根据四个参数和对象 cp，创建生成器 gernerator。

（5）设置 gernerator：

- 在 Model 中生成 dao 对象。
- 不生成字典文件。

（6）运行生成器，生成 Model、BaseModel 和 MappingKit 三类文件。

```
public class GeneratorModel {
    public static DataSource getDataSource() {
        PropKit.use("db.properties");
        DruidPlugin cp = new DruidPlugin(PropKit.get("url"), PropKit.get("user"),
            PropKit.get("password"));
```

```
            cp.start();
            return cp.getDataSource();
        }
        public static void main(String[] args) throws IOException {
            String baseModelPackageName = "cn.lrw.newsjfinal.model.base";
            String baseModelOutputDir = PathKit.getWebRootPath() + "/../src/cn/lrw/newsjfinal/
                model/base";
            String modelPackageName = "cn.lrw.newsjfinal.model";
            String modelOutputDir = baseModelOutputDir + "/..";
            Generator gernerator = new Generator(getDataSource(), baseModelPackageName,
                baseModelOutputDir, modelPackageName, modelOutputDir);
            gernerator.setGenerateDaoInModel(true);
            gernerator.setGenerateDataDictionary(false);
            gernerator.generate();
        }
    }
```

以 Java Application 运行（Run As）GeneratorModel 类，则会生成一系列 Model 相关文件，如图 6-7 所示。如果没有看到生成的文件、控制台也没有报错，只要刷新一下项目或项目中的 src，就能看到文件了。

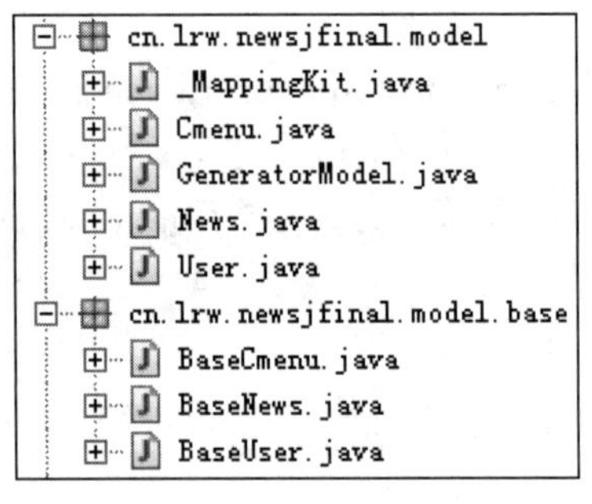

图 6-7　自动生成的 Model 相关文件

6.2.7　相关生成文件

从图 6-7 可以看出，在 cn.lrw.newsjfinal.model.base 包中，自动生成的是 BaseModel 一类的 Java 文件；在 cn.lrw.newsjfinal.model 包中，自动生成的是 Model 一类的 Java 文件。

BaseModel 是用于被最终的 Model 继承的基类，所有的 getter、setter 方法都将生成在此文件内，这样就保证了最终 Model 的清爽与干净。BaseModel 不需要人工维护，在数据库有任何变化时重新生成一次即可。

1. BaseModel 文件

以 BaseUser 类为例，可以看出类中仅仅定义了各属性的 getter 和 setter 方法。

```
@SuppressWarnings("serial")
public abstract class BaseUser<M extends BaseUser<M>> extends Model<M> implements IBean {
    public void setUid(java.lang.String uid) {
        set("uid", uid);
    }
```

```
        public java.lang.String getUid() {
            return getStr("uid");
        }
        public void setXm(java.lang.String xm) {
            set("xm", xm);
        }
        public java.lang.String getXm() {
            return getStr("xm");
        }
        public void setPwd(java.lang.String pwd) {
            set("pwd", pwd);
        }
        public java.lang.String getPwd() {
            return getStr("pwd");
        }
        public void setRole(java.lang.String role) {
            set("role", role);
        }
        public java.lang.String getRole() {
            return getStr("role");
        }
    }
```

2. Model 文件

以 User 类为例，可以看出它继承了 BaseUser，类中仅仅定义了静态 dao 对象。

```
@SuppressWarnings("serial")
public class User extends BaseUser<User> {
    public static final User dao = new User().dao();
}
```

3. MappingKit

MappingKit 用于生成 Table 到 Model 的映射关系，并且会生成主键/复合主键的配置，不需在 configPlugin(Plugins me)方法中书写任何样板式的映射代码。

```
public class _MappingKit {
    public static void mapping(ActiveRecordPlugin arp) {
        arp.addMapping("cmenu", "id", Cmenu.class);
        arp.addMapping("news", "id", News.class);
        arp.addMapping("user", "uid", User.class);
    }
```

4. DataDictionary 文件

DataDictionary 是指生成的数据字典，会生成数据表所有字段的名称、类型、长度、备注、是否主键等信息。当前项目已设置不生成这类文件。

6.2.8　创建 SysConfig 类

基于 JFinal 的 Web 项目需要创建一个继承自 JFinalConfig 类的子类，该类用于对整个 Web 项目进行配置。

1. 创建 SysConfig 类

在 cn.lrw.newsjfinal 包中创建 SysConfig 类，继承自 JFinalConfig，如图 6-8 所示。

图 6-8　创建 SysConfig 类

在 SysConfig 类中，需要实现 JFinalConfig 中的六个抽象方法，如图 6-9 所示。

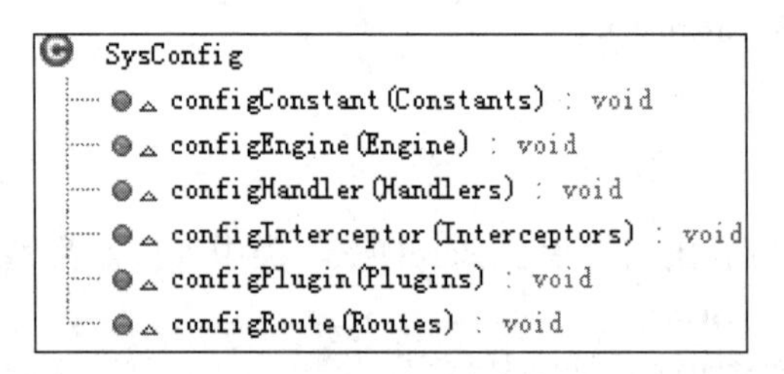

图 6-9　实现 JFinalConfig 中的六个抽象方法

2. 常量配置 configConstant

此方法用来配置 JFinal 常量值，如配置开发模式常量 devMode、Beetl 模板引擎。在开发模式（me.setDevMode(**true**)）下，JFinal 会对每次请求输出报告，如输出本次请求的 URL、Controller、Method 以及请求所携带的参数。以下还配置了前端页面使用 Beetl 模板引擎渲染输出。

```
@Override
public void configConstant(Constants me) {
    me.setDevMode(true);
    JFinal3BeetlRenderFactory rf = new JFinal3BeetlRenderFactory();
        rf.config();
        me.setRenderFactory(rf);
        GroupTemplate gt = rf.groupTemplate;
}
```

3. 路由配置 configRoute

此方法用来配置访问路由 Routes。Routes 类中有两个添加路由的方法：

（1）public Routes add(String controllerKey, Class<? extends Controller> controllerClass, String viewPath)

（2）public Routes add(String controllerKey, Class<? extends Controller> controllerClass)

- 参数 controllerKey 是指访问某个 Controller 所需要的一个字符串，该字符串唯一对应一个 Controller。controllerKey 仅能定位到 Controller。
- 参数 controllerClass 是该 controllerKey 所对应到的 Controller。
- 参数 viewPath 是指该 Controller 返回的视图的相对路径。当 viewPath 未指定时默认值为 controllerKey。

例如配置项 me.add("/news", NewsAct.class,"/WEB-INF/html")将 "/news" 映射到 NewsAct 这个控制器（Controller）；http://localhost:8080/newsjfinal/news/goList 将访问 NewsAct.goList() 方法；需要渲染 render 的页面必须存放在 WebRoot/WEB-INF/html 目录下。

```
@Override
public void configRoute(Routes me) {
    me.add("/", IndexAct.class, "/");
    me.add("/user", UserAct.class,"/WEB-INF/html");
    me.add("/news", NewsAct.class,"/WEB-INF/html");
    me.add("/menu", MenuAct.class);
    me.add("/file", FileAct.class);
}
```

4. 拦截器配置 configInterceptor

此方法用来配置 JFinal 的全局拦截器。全局拦截器将拦截所有 Action 请求，除非使用 @Clear 在 Controller 中进行清除。下述代码配置了名为 LoginInterceptor 的拦截器。

```
@Override
public void configInterceptor(Interceptors me) {
    me.add(new LoginInterceptor());
}
```

5. 插件配置 configPlugin

此方法用来配置 JFinal 的 Plugin，常用插件有 Druid 数据库连接池插件 DruidPlugin 和 ActiveRecord 数据库访问插件 ActiveRecordPlugin。通过以下的配置，可以在应用中使用 ActiveRecord 非常方便地操作数据库。

```
@Override
public void configPlugin(Plugins me) {
    PropKit.use("db.properties");
    DruidPlugin cp = new DruidPlugin(PropKit.get("url"), PropKit.get("user"), PropKit.get("password").
        trim());
    me.add(cp);
    ActiveRecordPlugin arp = new ActiveRecordPlugin(cp);
    _MappingKit.mapping(arp);
    me.add(arp);
    arp.setDialect(new MysqlDialect());
}
```

6. 模板配置 configEngine

此方法用来配置 Template Engine，向模板引擎中添加已经定义的 template function 模板文件，当前项目没有使用自定义的模板。

```
@Override
public void configEngine(Engine me) {
}
```

7. 处理器配置 configHandler

此方法用来配置 JFinal 的 Handler。Handler 可以接管所有 Web 请求，并对应用拥有完全的控制权，可以很方便地实现更高层的功能性扩展，当前项目没有使用自定义的处理器。

```
@Override
public void configHandler(Handlers me) {
}
```

6.2.9 配置 web.xml

打开 web.xml 文件，在 display-name 节点和 welcome-file-list 节点之间添加 JFinal 的 Filter，根据实际情况正确设置 JFinalFilter 的初始化参数值。

```
<filter>
    <filter-name>jfinal</filter-name>
    <filter-class>com.jfinal.core.JFinalFilter</filter-class>
    <init-param>
        <param-name>configClass</param-name>
        <param-value>cn.lrw.newsjfinal.SysConfig</param-value>
    </init-param>
</filter>
<filter-mapping>
    <filter-name>jfinal</filter-name>
    <url-pattern>/*</url-pattern>
</filter-mapping>
```

6.2.10 简单的首页

在 WebRoot 下创建一个基于默认模板的简单首页 index.html，页面 body 中的内容自定。

然后在 cn.lrw.newsjfinal.controller 包中创建 IndexAct 类，类中方法 render 实现系统首页 index.html。

```
@Clear
public class IndexAct extends Controller {
    public void index() {
        render("index.html");
    }
}
```

6.2.11 运行项目

由于在 SysConfig 类中配置的其他 Controller 类和 Interceptor 类还没有创建，因此先注释掉。然后运行基于 JFinal 框架的 JavaEE 项目，如果控制台 Console 没有报错信息，则在浏览器可以看到正常的首页信息。

6.2.12　考核任务

正确实现基于 JFinal 框架的入门项目。检查点：

（1）可以在浏览器看到默认首页 index.html 的内容，无乱码。（40 分）

（2）MyEclipse 的 Console 中没有报错（如 Exception）信息。（60 分）

6.3　日志系统

在 res 文件夹中创建日志配置文件 log4j.properties，配置内容与 3.4 节中相同，以便在调试时看到更多的日志。

6.4　公共方法类

在 cn.lrw.newsjfinal.utils 包中创建工具类 BaseUtil，封装一个用 Shiro 加密字符串的方法。

```
public class BaseUtil {
    public static String lrwCode(String password,String salt){
            if(salt==""){
                    salt="abcdefghijklmnopqrstuvwx";
            }
            return new Sha256Hash(password, salt, 1024).toBase64();
    }
}
```

6.5　创建业务逻辑类

通常要求所有业务逻辑写在 Service 中，所有 SQL 语句也只放在业务层 Service 中，或者放在外部 SQL 模板，用模板引擎管理，不要放在 Model 中，更不要放在 Controller 中。养成好习惯，有利于大型项目的开发与维护。所有的 DAO 对象也放在 Service 中，并且声明为 private。

6.5.1　UserSvc 类

在 cn.lrw.newsjfinal.service 包中创建 UserSvc 类，实现添加新用户、查询用户和用户数量等业务逻辑。

```
public class UserSvc {
    public static final UserSvc me = new UserSvc();
    private User dao = new User().dao();
    public void addU(User user) {
        user.save();
    }
    public User findU(String uid, String pwd) {
        return dao.findFirst("select * from user where uid = ? and pwd = ? ", uid, pwd );
```

```
    }
    public Long getCount() {
        return Db.queryLong("select count(*) from user");
    }
}
```

6.5.2 NewsSvc 类

在 cn.lrw.newsjfinal.service 包中创建 NewsSvc 类，实现新闻信息增删改查的业务逻辑。

```
public class NewsSvc {
    public static final NewsSvc me = new NewsSvc();
    private News dao = new News().dao();
    //===========增===============
    public void addNews(News news) throws Exception{
        news.save();
    }
    //===========删===============
    public void deleteNews(int id) throws Exception{
        News.dao.deleteById(id);
    }
    //===========改===============
    public void updateNews(News news) throws Exception{
        news.update();
    }
    //===========查===============
    //按新闻标题分页查询
    public Page<News> listDgNews(String title,int page,int rows){
        if(title == null || "".equals(title)) return News.dao.paginate(page, rows, "select *","from news
order by tjdate desc");
        else return News.dao.paginate(page, rows, "select *","from News news WHERE news.title
like ? order by news.tjdate desc", '%' +title+'%');
    }
    //按 id 查询/阅读新闻
    public News getNews(int id){
        News news=dao.findById(id);
        news.setHitnum(news.getHitnum()+1);//单击量增加
        news.update();
        return    news;
    }
    //查询新闻数量
    public int getNewsCount(){
        try{
            Long a=Db.queryLong("select count(*) from News");
            return    Integer.parseInt(a.toString());
        }catch(Exception e){
```

```
                e.printStackTrace();
                return 0;
            }
        }
    }
```

6.5.3　MenuSvc 类

在 cn.lrw.newsjfinal.service 包中创建 MenuSvc 类，实现 Tree 型菜单数据的获取。

```
public class MenuSvc {
    public static final MenuSvc me = new MenuSvc();
    private Cmenu dao = new Cmenu().dao();
    public List<Cmenu> listMenu(int pid){
        return dao.find("select * from cmenu where pid=?", pid);
    }
}
```

6.6　创建控制器类

完成以下 4 个控制器（Controller）类后，修改 SysConfig 类中的路由 Routes 配置，实现前端请求映射到相应控制器。

```
me.add("/user", UserAct.class,"/WEB-INF/html");
me.add("/news", NewsAct.class,"/WEB-INF/html");
me.add("/menu", MenuAct.class);
me.add("/file", FileAct.class);
```

6.6.1　UserAct 类

在 cn.lrw.newsjfinal.controller 包中创建 UserAct 类，实现用户登录 doLogin、退出 doLogout 和跳转到后台管理页面的请求 goIndex。

```
public class UserAct extends Controller {
    UserSvc userSvc=UserSvc.me;
    private Map<String,Object> jsonobj=new HashMap<String,Object>();
    @Clear
    public void doLogin(){
        String uid = getPara("uid");
        String pwd = getPara("pwd");
        try {
            Long n=userSvc.getCount();
            if(n==0){
                User user=new User();
                user.setUid("2953");
                user.setXm("lrw");
                user.setPwd(BaseUtil.lrwCode("123", ""));
                user.setRole("1");
                userSvc.addU(user);
```

```
            }
            pwd=BaseUtil.lrwCode(pwd, "");
            User user0 = userSvc.findU(uid, pwd);
            jsonobj.clear();
            if(user0 != null){
                jsonobj.put("ok", true);
                jsonobj.put("msg", "user/goIndex");
                setSessionAttr("me", user0);
            }else {
                jsonobj.put("ok", false);
                jsonobj.put("msg", "用户不存在");
            }
        } catch (Exception e) {
            jsonobj.put("ok", false);
            jsonobj.put("msg", "系统错误 2");
        }
        renderJson( jsonobj);
    }

    public void doLogout(){
        getSession().invalidate();
        redirect("/");
    }
    public void goIndex(){
        setAttr("me", (User)getSessionAttr("me"));
        render("admin.html");
    }
}
```

6.6.2 NewsAct 类

在 cn.lrw.newsjfinal.controller 包中创建 NewsAct 类，实现对新闻进行增删改查的请求。

```
public class NewsAct extends Controller {
    NewsSvc newsSvc=NewsSvc.me;
    private Map<String,Object> jsonobj=new HashMap<String,Object>();
```

1. 增（添加新闻）

```
//请求跳转到添加新闻页面
public void goAdd(){
    setAttr("me", (User)getSessionAttr("me"));
    render( "newsadd.html");
}
//保存添加的新闻
public void saveAdd(){
News news=getBean(News.class);
    try {
        news.setTjdate(new Date());   //提交日期由后端生成
        news.setHitnum(0);
```

```
            newsSvc.addNews(news);
            renderText( "ok");
        } catch (Exception e) {
            e.printStackTrace();
            renderText(   "系统错误 2");
        }
    }
```

2. 删（删除一条新闻）

```
    //删除一条新闻
    public void doDel1(){
    int id=getParaToInt("id");
        try{
            newsSvc.deleteNews(id);
            renderText( "ok");
        }catch(Exception e){
            e.printStackTrace();
            renderText(   "系统错误 2");
        }
    }
```

3. 改（修改一条新闻）

```
    //请求跳转到修改新闻页
    public void goEdit(){
    int id=getParaToInt("id");
    setAttr("news",newsSvc.getNews(id));
        render( "newsedit.html");
    }
    //保存修改后的新闻
    public void saveEdit(){
        News news=getBean(News.class);
        try {
            News news0=newsSvc.getNews(news.getId());
            news0.setContent(news.getContent());
            news0.setCruser(news.getCruser());
            news0.setTitle(news.getTitle());
            newsSvc.updateNews(news0);
            renderText( "ok");
        } catch (Exception e) {
            e.printStackTrace();
            renderText(   "系统错误 2");
        }
    }
```

4. 查（根据条件查找新闻）

```
        //请求跳转到后台新闻列表页
        public void goList(){
            render( "newslist.html");
        }
```

```
    //统计新闻总数量
    @Clear
    public void getCount(){
        int c=0;
        try{
            c=newsSvc.getNewsCount();
        }catch(Exception e){
            e.printStackTrace();
            c=0;
        }
        jsonobj.clear();
        jsonobj.put("newscount", c);
        renderJson( jsonobj);
    }
    //查询/阅读一条新闻
    @Clear
    public void getNews(){
    int id=getParaToInt("id");
    setAttr("news",newsSvc.getNews(id));
        render( "newsread.html");
    }
    //分页查询新闻
    @Clear
    public void listNews(){
        String s_name=getPara("s_name");
        int page=getParaToInt("page");
        int rows=getParaToInt("rows");
        Page<News> newslist=newsSvc.listDgNews(s_name,page,rows);
        jsonobj.clear();
        jsonobj.put("total", newslist.getTotalRow());
        jsonobj.put("rows", newslist.getList());
        renderJson( jsonobj);
    }
}
```

6.6.3 MenuAct 类

在 cn.lrw.newsjfinal.controller 包中创建 MenuAct 类，实现对菜单数据的封装。

```
public class MenuAct extends Controller {
    MenuSvc menuSvc=MenuSvc.me;
    public void menutree() {
        User user = (User) getSessionAttr("me");
        String role=user.getRole();
        //=====================一级菜单==========================
        List<Cmenu> menulist=menuSvc.listMenu(0);
        List<EasyUITree> eList = new ArrayList<EasyUITree>();
        if(menulist.size() != 0){
```

```
                for (int i = 0; i < menulist.size(); i++) {
                    Cmenu t = menulist.get(i);
                    if(!t.getPermission().contains(role))continue;
                    EasyUITree e = new EasyUITree();
                    e.setId(t.getId()+"");
                    e.setText(t.getName());
                    List<EasyUITree> eList2 = new ArrayList<EasyUITree>();
                //====================二级菜单===========================
                    List<Cmenu> menu2 = menuSvc.listMenu(t.getId());
                    for (int j = 0; j < menu2.size(); j++) {//二级菜单
                        Cmenu t2 = menu2.get(j);
                        if(!t2.getPermission().contains(role))continue;
                        Map<String,Object> attributes = new HashMap<String, Object>();
                        attributes.put("url", t2.getUrl());
                        attributes.put("role", t2.getPermission());
                        EasyUITree e1 = new EasyUITree();
                        e1.setAttributes(attributes);
                        e1.setId(t2.getId()+"");
                        e1.setText(t2.getName());
                        e1.setState("open");
                        eList2.add(e1);
                    }
                    e.setChildren(eList2);
                    e.setState("closed");
                    eList.add(e);
                }
            }
            renderJson( eList);
        }
    }
```

6.6.4　FileAct 类

在 cn.lrw.newsjfinal.controller 包中创建 FileAct 类，实现前端页面使用百度在线编辑器 UEditor 上传文件时，后端接收并写入磁盘。

```
    @Clear
    public class FileAct extends Controller {
        public void bdupfile()throws IOException{
            HttpServletResponse res=getResponse();
            HttpServletRequest req=getRequest();
            req.setCharacterEncoding( "UTF-8" );
            res.setHeader("Content-Type" , "text/html");
            String rootPath = PathKit.getWebRootPath();
            renderHtml(new ActionEnter( req, rootPath ).exec());
        }
    }
```

6.7 前端页面

由于当前项目中的前端页面均是使用 Beetl 模板渲染的 html 页面，与基于 Nutz 的项目类似，但也有些小的区别。所以可以将第 3 章的基于 Nutz 项目中 include 和 error 里面的文件全部分别复制到当前项目的 include 和 error 中，只需进行少量的修改，即可用于当前项目。

当前项目仍然以开发简易的新闻发布系统为例，旨在了解和实践 JFinal 项目的搭建和实现，了解项目的工作流程。前端页面的主要内容与第 3 章的基于 Nutz 项目的页面内容基本一样，只需稍作修改或调整。

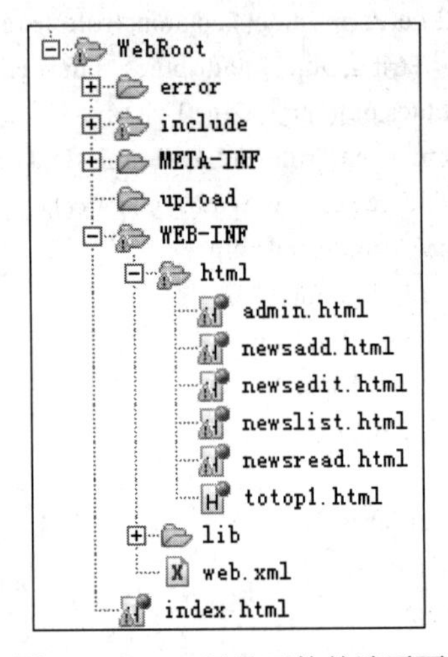

图 6-10 html 目录下的前端页面

6.7.1 系统首页

index.html 是系统首页，分页显示新闻列表和登录窗口。首页位于 WebRoot 目录下，可以直接访问，所以可以使用相对路径引用 JS、CSS 文件和图片文件。详细的实现方法，请参照 3.5.1 美化系统首页、3.5.2Ajax 方法、3.5.3 更友好的 alert、3.5.4 标题图标、3.8.10.1 前台新闻列表。

页面 index.html 内容如下：

```
<!DOCTYPE html PUBLIC "-//W3C//DTD XHTML 1.0 Transitional//EN" "http://www.w3.org/TR/xhtml1/DTD/xhtml1-transitional.dtd">
<html xmlns="http://www.w3.org/1999/xhtml">
<head>
    <meta http-equiv="Content-Type" content="text/html; charset=UTF-8" />
    <title>简易新闻发布系统</title>
    <link rel="shortcut icon" type="image/x-icon" href="./include/img/logo.png" />
    <link rel="stylesheet" type="text/css" href="./include/css/main.css" />
    <link rel="stylesheet" type="text/css" href="./include/easyui/themes/default/easyui.css" />
    <link rel="stylesheet" type="text/css" href="./include/easyui/themes/icon.css" />
    <script type="text/javascript" src="./include/js/jquery.min.js"></script>
```

```
        <script type="text/javascript" src="./include/easyui/jquery.easyui.min.js"></script>
        <script type="text/javascript" src="./include/easyui/locale/easyui-lang-zh_CN.js"></script>
        <script type="text/javascript" src="./include/js/login.js"></script>
</head>
<body>
<div style="float:right;padding-right:20px;">
<a id="a" href="#" style="margin-right:15px;" >登录</a>      <a id="b" href="#" >新闻</a>
</div>
<div class="login">
    <div id="llogin" class="box png">
        <div class="logo png"></div>
        <div class="input">
            <div class="log" id="login_form">
                <div class="name">
                    <label>用户名</label><input type="text" class="text" id="uid" placeholder=
                        "用户名"   tabindex="1" />
                </div>
                <div class="pwd">
                    <label>密   码</label><input type="password" class="text" id="pwd"
                        placeholder="密码"   tabindex="2" />
                    <input id="login_submit" type="button" class="submit" tabindex="3" value="登录" />
                    <div class="check"></div>
                </div>
                <div class="tip"></div>
            </div>
        </div>
    </div>
    <div id="lnews" class="l-wrap">
    <div>
        <div>
            <div class="l-news">
                <div class="nheader">
                    <table cellspacing="0" cellpadding="0"><tbody>
                        <tr><td><h3>通知新闻：</h3></td></tr>
                    </tbody></table>
                </div>
                <div class="nlist">
                    <table id="newstable" width="100%">
                      <tbody>
                        <tr id="trpp"><td colspan="3" align="left"> </td></tr>
                    </tbody></table>
                </div>
                <div id="pp" style="background:#efefef;"></div>
            </div>
        </div>
    </div>
```

```
        </div>
        <div class="air-balloon ab-1 png"></div>
        <div class="air-balloon ab-2 png"></div>
        <div class="footer"></div>
    </div>
    </body>
    </html>
```

页面中引用的 login.js，实现登录窗口、新闻列表的分页动态加载和气球的动态运动，代码如下。

```
var base="./";
var pageN=1,pageTotal=100;
$(function(){
    $('#login_form input').keydown(function (e) {
        if (e.keyCode == 13)
        {
            checkUserName();
        }
    });
    $("#login_submit").click(checkUserName);
    $.ajax({
        url:"./news/getCount",
        type:"post",
        success: function(res){
            pageTotal=parseInt(res);
            listNews(1,10);loadPager();
        },
        error:function(res){
            $.messager.alert("系统提示","系统错误","error");
        }
    });
    $("#llogin").hide();
    $("#a").click(function(){
    $("#llogin").show();
    $("#lnews").hide();
    });
    $("#b").click(function(){
    $("#llogin").hide();
    $("#lnews").show();
    });
});
function listNews(pageNumber,pageSize){
    $.ajax({
        url:"./news/listNews",
        data:{"page":pageNumber,"rows":pageSize},
        type:"post",
        success: function(res){
```

```
                $(".inews").remove();
                res=JSON.parse(res);
                if(res.total<=0){
                    var tr="<tr class='inews' height=\"25\"><td >";
                        tr+="<div class='t'>暂无相关新闻</div>";
                        tr+="</td><td width='1%' nowrap=''><span > </span></td></tr>";
                    $("#trpp").before(tr);
                }
                else {
                    pageN=pageNumber;
                    pageTotal=res.total;
                    var rows=res.rows;
                    for(var i=0;i<rows.length;){
                        var row=rows[i];
                        var tr="<tr class='inews' height=\"25\"><td >";
                            tr+="<div class='t'><a href='./news/getNews?id="+row.id+"' target='_blank'
                                title='"+row.title+"'>"+row.title+"</a></div>";
                            tr+="</td><td width='1%' nowrap=''><span >"+row.tjdate.substr(0,10)+
                                " </span></td></tr>";
                        $("#trpp").before(tr);
                        i++;
                        if(i%5==0){
                            tr="<tr class='inews' height='1'><td colspan='2' align='center'>";
                            tr+="<hr style='border-style:dashed;border-color:#999999;width:99%;
                              height:1px;border-width:1px 0px 0px 0px;visibility:inherit'></td></tr>";
                            $("#trpp").before(tr);
                        }
                    }
                }
            },
            error:function(res){
                $.messager.alert("系统提示","系统错误","error");
            }
        })
}
function loadPager(){
    $('#pp').pagination({
        total:pageTotal,
        pageSize:10,
        pageNumber:pageN,
        displayMsg:'{from}/{to}  共{total}条',
        onSelectPage:function(pageNumber, pageSize){
        listNews(pageNumber,pageSize);
        }
    });
}
```

```
function checkUserName()//登录前，校验用户信息
{
        var a=$('#uid').val();
        var b=$('#pwd').val();
        if(a==""){
        alert("请输入用户名");return;
        $.messager.alert('系统提示',"请输入用户名","warning");
        return;
        }
        if(b==""){$.messager.alert('系统提示',"请输入登录密码","warning");return;}
        $.ajax({
                url : base+"user/doLogin",
        //只封装和传输指定的数据
                data :{"uid":a,"pwd":b},
                type:"POST",
                success : function (res) {
                        if (res.ok) {
                                window.location.href=base+res.msg;
                        }else {$.messager.alert('系统提示',res.msg,"warning");                }
                        return false;
                },
                error : function(res) {$.messager.alert('系统提示',"系统错误！","error");        }
        });
}
//============原代码，作用是显示动态效果================
$(function(){
      airBalloon('div.air-balloon');
});
function airBalloon(balloon){
      var viewSize = [] , viewWidth = 0 , viewHeight = 0 ;
      resize();
      $(balloon).each(function(){
            $(this).css({top: rand(40, viewHeight * 0.5 ) , left : rand( 10 , viewWidth - $(this).width() ) });
            fly(this);
      });
      $(window).resize(function(){
            resize()
            $(balloon).each(function(){
                  $(this).stop().animate({top: rand(40, viewHeight * 0.5 ) , left : rand( 10 , viewWidth -
                      $(this).width() ) } ,1000 , function(){
                       fly(this);
                  });
            });
      });
      function resize(){
            viewSize = getViewSize();
```

```
            viewWidth = $(document).width() ;
            viewHeight = viewSize[1] ;
        }
        function fly(obj){
            var $obj = $(obj);
            var currentTop = parseInt($obj.css('top'));
            var currentLeft = parseInt($obj.css('left') );
            var targetLeft = rand( 10 , viewWidth - $obj.width() );
            var targetTop = rand(40, viewHeight /2 );
            /*求两点之间的距离*/
            var removing = Math.sqrt( Math.pow( targetLeft - currentLeft , 2 ) + Math.pow( targetTop -
                currentTop , 2 ) );
            /*每秒移动 24px ， 计算所需要的时间，从而保持 气球的速度恒定*/
            var moveTime = removing / 24;
            $obj.animate({ top : targetTop , left : targetLeft} , moveTime * 1000 , function(){
                setTimeout(function(){
                    fly(obj);
                }, rand(1000, 3000) );
            });
        }
        function rand(mi,ma){
            var range = ma - mi;
            var out = mi + Math.round( Math.random() * range) ;
            return parseInt(out);
        }
        function getViewSize(){
            var de=document.documentElement;
            var db=document.body;
            var viewW=de.clientWidth==0 ?  db.clientWidth : de.clientWidth;
            var viewH=de.clientHeight==0 ?  db.clientHeight : de.clientHeight;
            return Array(viewW,viewH);
        }
    };
```

6.7.2　出错跳转页

当出现系统错误时，直接跳转到 nologin.html 页。

```
<!DOCTYPE HTML>
<html>
<head>
<meta charset="UTF-8"/>
<link rel="icon" href="../include/img/logo.png" type="image/x-icon" />
<link href='../include/css/404.css?family=Love+Ya+Like+A+Sister' rel='stylesheet' type='text/css'>
<script type="text/javascript" src="../include/js/jquery.min.js"></script>
<title>无效访问</title>
</head>
<body>
```

```
<div class="wrap">
    <div class="logo">
        <div class="errcode"><span>out</span></div>
        <p>对不起，您没有登录或者登录已超时!!!</p>
        <div class="sub">
            <p><a href="###"  id="index">返回首页</a></p>
        </div>
    </div>
</div>
<script type="text/javascript">
$("#index").click(function() { reDo(); });
setTimeout(reDo, 5000);//
function reDo(){ top.location.href = "../";   }
</script>
</body>
</html>
```

6.7.3 新闻阅读页

在新闻阅读页 newsread.html 中使用了 Beetl 引擎，应注意${ctxPath}的应用。

```
<!DOCTYPE html>
<html>
<head>
<meta charset="UTF-8">
<title>读新闻</title>
<link rel="shortcut icon" href="${ctxPath}/include/img/logo.png" type="image/x-icon" />
<link rel="stylesheet" type="text/css" href="${ctxPath}/include/easyui/themes/default/easyui.css">
<link rel="stylesheet" type="text/css" href="${ctxPath}/include/easyui/themes/icon.css">
<link rel="stylesheet" type="text/css" href="${ctxPath}/include/css/news.css">
<script type="text/javascript" src="${ctxPath}/include/js/jquery.min.js"></script>
</head>
<body>
    <div style="background:#B3DFDA;padding:0 10px 0 10px;vertical-align: middle;">
        <img src="${ctxPath}/include/img/logo.png" width="126" height="50" />
        <div style="float:right;line-height:50px;margin-right:10px;font-size: 9pt;">
            <span>【</span><a style="color:blue;" href="javascript:window.close();"><span>关闭
                窗口</span></a><span>】</span>
        </div>
    </div>
<div class="ndetail">
    <div class="ntitle">${news.title}</div>
    <div class="nauthor">
        <div>来源:  <strong>${news.cruser}</strong>   发布时间: 
        <strong>${news.tjdate,dateFormat="yyyy-MM-dd"}</strong>  访问量: 
        <strong>[<span>${news.hitnum}</span>]</strong></div>
    </div>
    <div class="nbody">
```

```
        <div id="vsb_content"> ${news.content}</div>
    </div>
</div>
<% include("totop1.html"){}%>
</body>
</html>
```

6.7.4 后台 Layout

新闻管理的后台布局页 admin.html 主要实现内容的布局、树形菜单的加载和主页的加载。

1. 引用文件

```
<link rel="shortcut icon" href="${ctxPath}/include/img/logo.png">
<link rel="stylesheet" type="text/css" href="${ctxPath}/include/easyui/themes/default/easyui.css">
<link rel="stylesheet" type="text/css" href="${ctxPath}/include/easyui/themes/icon.css">

<script type="text/javascript" src="${ctxPath}/include/js/jquery.min.js"></script>
<script type="text/javascript" src="${ctxPath}/include/easyui/jquery.easyui.min.js"></script>
<script type="text/javascript" src="${ctxPath}/include/easyui/locale/easyui-lang-zh_CN.js"></script>
<script type="text/javascript" src="${ctxPath}/include/js/lrwtab.js"></script>
```

2. 页面布局

```
<body class="easyui-layout">
    <div data-options="region:'north',border:false"
        style="background:#fff;padding:0 10px 0 10px;vertical-align: middle;">
        <img src="${ctxPath}/include/img/logo.png" width="126" height="50" />
        <div style="float:right;line-height:50px;margin-right:10px;">
            <a id="logout" href="#" class="easyui-linkbutton"
                data-options="iconCls:'icon-cancel'">退出</a>
        </div>
        <div style="float:right;line-height:50px;margin-right:10px;">登录用户：${me.xm}
            |</div>
    </div>
    <div data-options="region:'west',split:true,title:'系统菜单'"
        style="width:180px;padding:10px;">
        <ul id="menutree" class="easyui-tree"></ul>
    </div>
    <div data-options="region:'south',border:false"
        style="height:30px;padding:5px;text-align:center;font-family: arial;
            color: #A0B1BB;">Copyright
        © 2017 JavaEE. All Rights Reserved.</div>
    <div data-options="region:'center'">
        <div id="tabs" class="easyui-tabs" fit="true" border="false"></div>
    </div>
</body>
```

3. 加载树形菜单

```
<script>
var opened_node;
```

```
$(function(){
$("#menutree").tree(
{
    url : "${ctxPath}/menu/menutree",
    animate : true,
    onClick : function(node) {
        if (!node.attributes) {
            if (!opened_node) {
                $("#menutree").tree('expand', node.target);
                opened_node = node.target;
            } else if (opened_node != node.target) {
                $("#menutree").tree('collapse', opened_node);
                $("#menutree").tree('expand', node.target);
                opened_node = node.target;
            }
        } else {
            swNewTab(node.text,"${ctxPath}" +node.attributes.url);
        }
    },
    onLoadSuccess : function(node, data) {
        $("#menutree").tree('expandAll');
    }
});
$("#logout").click(function(){
    top.location.href="${ctxPath}/user/doLogout";
});
swNewTab('新闻管理',"${ctxPath}/news/goList");
});
</script>
```

6.7.5 新闻列表页

新闻列表页 newslist.html 页面上使用 datagrid 加载分页新闻，每条新闻后有修改和删除的链接，工具栏中有标题关键字输入框、查询按钮。

1. 引用文件

```
<link rel="shortcut icon" href="${ctxPath}/include/img/logo.png">
    <link rel="stylesheet" type="text/css" href="${ctxPath}/include/easyui/themes/default/easyui.css">
    <link rel="stylesheet" type="text/css" href="${ctxPath}/include/easyui/themes/icon.css">
    <script src="${ctxPath}/include/js/jquery.min.js"></script>
    <script type="text/javascript" src="${ctxPath}/include/easyui/jquery.easyui.min.js"></script>
<script type="text/javascript" src="${ctxPath}/include/easyui/locale/easyui-lang-zh_CN.js"></script>
```

2. 页面设计

```
<body>
 <table id="dg" cellpadding="2" ></table>
<div id="tb" style="padding:5px;">
     <input id="s_name" class="easyui-textbox"   data-options="prompt:'标题关键字...'" style=
```

```
        "width:200px;height:32px">
    <a id="s_news" href="#" class="easyui-linkbutton" data-options="iconCls:'icon-search'">查询</a>
</div>
</body>
```

3. 加载新闻列表

```
<script>
var s_name="",id="",title="";
function loadGrid(){
    s_name=$("#s_name").textbox("getValue");
    $("#dg").datagrid({
        width:800,height:500,nowrap:false,
        striped:true,border:true,collapsible:false,
        url:"${ctxPath}/news/listNews",
        queryParams:{"s_name":s_name},
        pagination:true,
        rownumbers:true,
        fitColumns:true,pageSize:20,
        loadMsg:'数据加载中...' ,
        columns:[[
            {title:'标题', field:'title',width:200,formatter: function(value,row,index){
                return '<span style="white-space: nowrap;" title='+value+'>'+
                    (value?value:'')+ '</span>';
            }},
            {title:'发布时间', field:'tjdate',width:100},
            {title:'操作', field:'hitnum',width:100, formatter: function(value,row,index){
                var p="<a href=\"javascript:editNews('"+row.id+"')\">修改</a>";
                p+="|<a href=\"javascript:delNews('"+ row.id+ "','"+ row.title+ "')\">
                    删除</a>";
                return p;
            }}
        ]],
        toolbar:'#tb'
    });
}
function editNews(id){
    parent.swNewTab("修改新闻信息","${ctxPath}/news/goEdit?id="+id);
}
function delNews(newsid,title0){
    id=newsid;title=title0;
    parent.$.messager.confirm("系统提示", "您确认要删除“"+title+"”吗？ ", function(r){
      if (r){
        $.ajax({
            url:"${ctxPath}/news/doDel1",
            data:{"id":id},
            type:"post",
            success: function(res){
```

```
                    if(res=="ok"){
                        parent.$.messager.alert("系统提示","您已删除新闻："+title,"info");
                        id="";s_name="";
                        loadGrid();
                    }else {
                        parent.$.messager.alert("系统提示",res,"error");
                    }
                    return false;
                },
                error:function(res){
                    parent.$.messager.alert("系统提示","系统错误","error");
                }
            })
        }
    });
}

$(function(){
    loadGrid();
    $("#s_news").click(function(){
        loadGrid();
    });
     $("#tb").bind("keydown",function(e){
        var theEvent = e || window.event;
        var code = theEvent.keyCode || theEvent.which || theEvent.charCode;
        if (code == 13) {
            loadGrid();
        }
    });
})
</script>
```

6.7.6 新闻添加页

新闻添加页 newsadd.html 页面，使用百度在线编辑器 UEditor。

1. 文件的引用

```
<link rel="shortcut icon" href="${ctxPath}/include/img/logo.png">
<link rel="stylesheet" type="text/css" href="${ctxPath}/include/easyui/themes/default/easyui.css">
<link rel="stylesheet" type="text/css" href="${ctxPath}/include/easyui/themes/icon.css">
<script src="${ctxPath}/include/js/jquery.min.js"></script>
<script type="text/javascript" src="${ctxPath}/include/easyui/jquery.easyui.min.js"></script>
<script type="text/javascript" src="${ctxPath}/include/easyui/locale/easyui-lang-zh_CN.js"> </script>
<script type="text/javascript" charset="UTF-8" src="${ctxPath}/include/ueditor/ueditor.config.js">
</script>
<script type="text/javascript" charset="UTF-8" src="${ctxPath}/include/ueditor/ueditor.all.min.js">
</script>
<script type="text/javascript" charset="UTF-8" src="${ctxPath}/include/ueditor/lang/zh-cn/zh-cn.js"> </script>
```

2. 页面设计

```
<body>
    <div class="easyui-panel" style="padding:5px 2px">
        <form>
            <table cellpadding="5">
                <tr>
                    <td style="width:100px;">新闻标题：</td>
                    <td style="width:880px;"><input id="title" class="easyui-textbox" data-options=
                    "prompt:'新闻标题',required:true" style="width:90%;height:32px"></td>
                </tr>
                <tr>
                    <td>新闻发布者：</td>
                    <td><input id="cruser" class="easyui-textbox" value="${me.xm}" data-options=
                    "prompt:'发布人',required:true" style="width:90%;height:32px"></td>
                </tr>
                <tr>
                    <td style="vertical-align: top;">新闻内容：</td>
                    <td><script id="content" type="text/plain" style="width:89%;height:300px;">
                    </script></td>
                </tr>
            </table>
        </form>
        <div style="text-align:center;">
            <a id="savenews" href="#" class="easyui-linkbutton" iconCls="icon-ok" style=
            "width:132px;height:32px">保存</a>
        </div>
    </div>
</body>
```

3. 发布新闻

```
<script type="text/javascript">
var ue;
$(function() {
    ue = UE.getEditor('content');
    $('#savenews').click(function() {   //发布新闻前，要校验
        var a = $("#title").textbox("getValue");
        var b = ue.getContent();
        var c = $("#cruser").textbox("getValue");
        if (a.length <= 0) {
            $.messager.alert("系统提示", "必须填写新闻标题", "warning");
            return;
        }
        if (b.length <= 0) {
            $.messager.alert("系统提示", "必须填写新闻内容", "warning");
            return;
        }
        if (c.length <= 0) {
```

```
                $.messager.alert("系统提示", "必须填写发布人姓名或者发布机构名称", "warning");
                return;
            }
            $.ajax({
                type : 'POST',
                url : "${ctxPath}/news/saveAdd",
                data : {"news.title" : a,"news.content" : b,"news.cruser" : c},
                success : function(res) {
                    if (res == "ok") {
                        parent.$.messager.alert("系统提示", "你已添加新闻:"+ $("#title").val(), "info");
                    } else {
                        parent.$.messager.alert("系统提示", "添加失败！","error");
                    }
                    return false;
                },
                error : function(res) {
                    parent.$.messager.alert("系统提示", "系统错误！", "error");
                }
            });
        });
    });
    </script>
```

6.7.7 新闻修改页

新闻修改页 newsedit.html 页面，修改内容与新闻添加页类似，引用一样的文件。

1. 页面设计

```
    <body>
          <div class="easyui-panel" style="padding:5px 2px">
              <form>
                <table cellpadding="5">
                <tr><td style="width:100px;">新闻标题：</td><td style="width:880px;">
                      <input id="title" class="easyui-textbox" data-options="prompt:'新闻标题',required:
                      true" style="width:90%;height:32px">
                </td></tr>
                <tr><td>新闻发布者：</td><td>
                      <input id="cruser" class="easyui-textbox" data-options="prompt:'发布人',required:
                      true" style="width:90%;height:32px">
                </td></tr>
                <tr><td style="vertical-align: top;">新闻内容：</td><td>
                      <script id="content" type="text/plain" style="width:89%;height:300px;"></script>
                </td></tr>
                </table>
              </form>
              <div style="text-align:center;">
                  <a id="savenews" href="#" class="easyui-linkbutton" iconCls="icon-ok" style=
                  "width:132px;height:32px">保存</a>
```

```
            </div>
        </div>
    </body>
```

2. 保存修改后的新闻

```
    <script type="text/javascript">
    var ue;
    $(function(){
        ue = UE.getEditor('content');
        $("#title").textbox("setValue","${news.title}");
        ue.ready(function() {
            ue.setContent("");
            ue.execCommand('insertHtml', '${news.content}');
        });
        $("#cruser").textbox("setValue","${news.cruser}");
        $('#savenews').click(function(){//发布新闻前，要校验
            var a=$("#title").textbox("getValue");
            var b=ue.getContent();
            var c=$("#cruser").textbox("getValue");
                if(a.length<=0){$.messager.alert("系统提示","必须填写新闻标题","warning");return;}
                if(b.length<=0){$.messager.alert("系统提示","必须填写新闻内容","warning");return;}
                if(c.length<=0){$.messager.alert("系统提示","必须填写发布人姓名或者发布机构
                名称","warning");return;}
                $.ajax({
                        type: 'POST',
                        url : "${ctxPath}/news/saveEdit",
                        data : {"news.title":a,"news.content":b,"news.cruser":c,"news.id":${news.id}},
                        success : function (res) {
                                if(res=="ok"){
                                        parent.$.messager.alert("系统提示","你已修改新闻:"+
                                        $("#title").val(), "info");
                                }else{
                                        parent.$.messager.alert("系统提示","修改失败！ ","error");
                                }
                                return false;
                        },
                        error : function(res) {parent.$.messager.alert("系统提示","系统错误！ ", "error");}
                });
        });

    });
    </script>
```

6.8　增强安全

可以在 cn.lrw.newsjfinal.interceptor 包中创建一个自定义的拦截器，实现对某些请求不过

滤、某些请求必须过滤、过滤检测不通过时跳转到指定页面。

6.8.1 拦截器 LoginInterceptor

拦截未登录的用户，自动跳转到 nologin.html 页面。

```
public class LoginInterceptor implements Interceptor {
    @Override
    public void intercept(Invocation inv) {
        Controller controller = inv.getController();
        User user = (User) controller.getSessionAttr("me");
        if (user == null) {
            controller.redirect("/error/nologin.html");
            return;
        } else {
            inv.invoke();
        }
    }
}
```

6.8.2 配置拦截器

需要在 SysConfig 类和 Controller 类中进行配置，使得拦截器发挥作用。

1. 在 SysConfig 类配置拦截器

```
@Override
public void configInterceptor(Interceptors me) {
    me.add(new LoginInterceptor());
}
```

2. 添加拦截器

在需要拦截器起作用的 Controller 类的类名前添加拦截器注解@Before(LoginInterceptor.class)。

```
@Before(LoginInterceptor.class)
public class UserAct extends Controller
```

3. 取消拦截

在不需要拦截的方法前添加@Clear，取消拦截器的拦截作用。

```
@Clear
public void doLogin()
```

6.9 考核任务

正确实现基于 JFinal 框架的新闻信息增删改查，添加与修改新闻可以上传图片等功能。

（1）添加新闻。（20 分）

（2）修改新闻。（20 分）

（3）后台新闻列表。（20 分）

（4）删除新闻。（20 分）

（5）前台新闻列表。（10 分）

（6）阅读新闻。（10 分）

本章小结

本章开发基于 JFinal 框架的 JavaEE 项目，编码量少、配置文件少、配置内容少、引入的 Jar 包少，所以可以快速完成项目的开发。比主流框架 JFinal 更简单，但主流框架有商业机构支持，所以用户群很庞大。

JFinal 框架在开源中国的开源软件里面排名靠前，也就是说，有不少的软件开发人员在使用它，认可它的优点，因为简单，能在很短的时间里面学会并开发出 JavaEE 的项目。

第 7 章　项目部署

实现项目的部署，首先需要将开发机器上的数据库导出；然后在部署项目的服务器上重建数据库；最后将项目文件放在服务器的容器中，如 Tomcat 的 webapps，如有必要，可能要修改一些配置文件的参数，例如数据库的连接。

7.1　数据库的导出

以 dbnews1 数据库为例，假定借助 HeidiSQL 可视化工具实现数据的导出，操作步骤如下：

（1）用 HeidiSQL 工具打开本机的 dbnews1 数据库。

（2）右键单击数据库名称 dbnews1，在弹出的快捷菜单中选择 Export database as SQL 命令。

（3）在打开的 Table tools 对话框中，勾选 Database 的 Drop 和 Create、勾选 Table 的 Drop 和 Create、选择 Data 中的 Insert、选择 Output 的 Single .sql file，最后选择 SQL 文件保存的路径和文件名，假定文件名为 dbnews1.sql，单击 Export 按钮，如图 7-1 所示。

（4）等待所有表的导出状态显示 100%，且没有报错信息，就完成了整个数据库中所有表的结构及数据的导出。

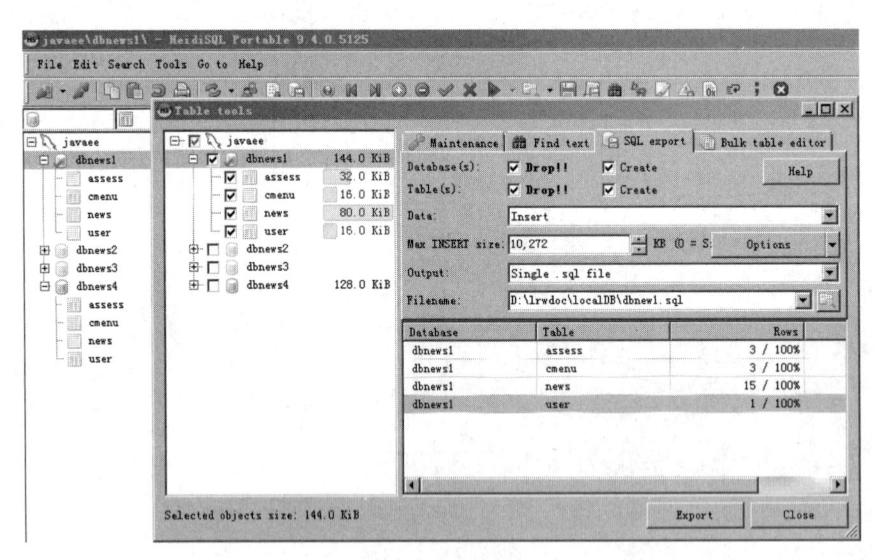

图 7-1　导出数据库

7.2　数据库的导入

假定服务器上已经安装好 MySQL，并且已经运行。导入数据的步骤如下：

（1）用 HeidiSQL 工具打开服务器的 MySQL 数据库。

（2）选择 File 中的 Load SQL file 菜单项或单击工具栏中的 Load SQL file 按钮，准备加载 SQL 文件。如果 SQL 文件太大，可以直接选择 File 中的 Run SQL file 菜单命令。

（3）选择 7.1 节中导出的 SQL 文件，如 dbnews1.sql。

（4）SQL 文件加载完成后，单击工具栏中的 Execute SQL 按钮，如图 7-2 所示，把整个数据库 dbnews1 部署到服务器上。

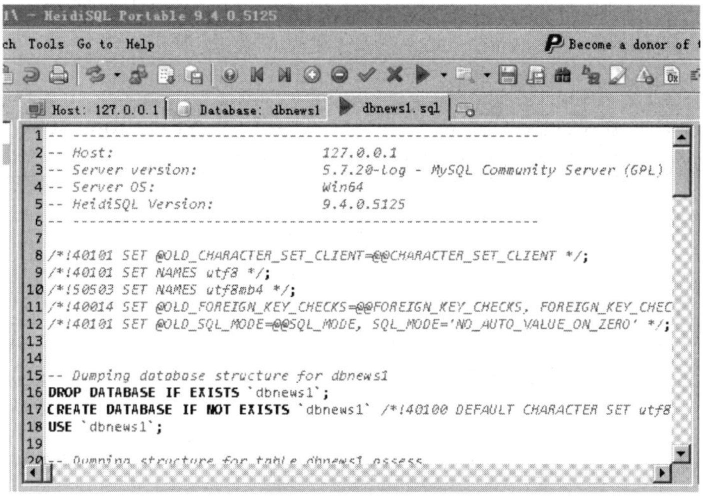

图 7-2　导入数据库

注意：如果导出的文件包含较多的数据，导入时会提示不能加载，但可以直接运行；如果部分表中的数据特别多，则可以将该表单独导出成一个 SQL 文件，即将数据库导出成若干个无重复表数据的 SQL 文件。

7.3　项目导出与部署

如果用来部署项目的服务器与开发项目所用的计算机运行 JavaEE 项目的环境一样，如数据库的端口配置一样，且不需要把测试数据迁移到服务器，建议采用导出 war 文件的方法，操作步骤简单方便。

（1）在 MyEclipse 中，选中需要导出的项目，如 newsnutz。

（2）选择 File 菜单中的 Export 命令。

（3）导出类型选择 war file，假定导出文件为 newsnutz.war。

（4）把 newsnutz.war 部署到服务器的 webapps 中，启动 Tomcat（如果此前 Tomcat 正在运行，则先将它停止，再启动）。

（5）启动完成后，在 webapps 中会自动生成 newsnutz 项目文件夹。

7.4　项目复制与部署

如果需要把测试数据（比如上传的文件）迁移到服务器，服务器上数据库的配置与开发机器不相同，建议采用复制项目文件夹的方法。

假定 MyEclipse 中指定的服务器为自己安装的 Tomcat8，找到 Tomcat8 安装目录下的 webapps，比如 D:\jtmlrw\tomcat8\webapps，可以看到项目文件夹，如 newsnutz。

（1）停止开发机器上 Tomcat8 的运行，复制项目文件夹 newsnutz。

（2）停止服务器上 Tomcat 的运行，在服务器部署项目的文件夹中粘贴复制得到的项目文件夹 newsnutz。

（3）如果配置有变化，则修改配置文件中的参数。如基于 Nutz 框架的项目，数据库配置不相同，则修改\WEB-INF\classes\ioc\dao.js 中的数据库配置参数。

（4）重新启动服务器上的 Tomcat。

（5）如果数据库的数据已导入、项目文件已部署到服务器，Tomcat 服务器运行后，检查测试项目的运行状况、项目中各功能模块的使用。有问题时，查明原因，立即解决。

7.5 考核任务

选择完成质量最好的一个完整项目上传到服务器。

（1）导出项目所使用的数据库为单一 SQL 文件（含数据库和表的创建、表结构、表中的数据）。（50 分）

（2）正确部署和展示项目，尤其注意配置文件中相关参数的正确配置，如数据库的配置。（50 分）

本章小结

本章介绍了项目的部署。把开发环境中完成的项目部署到服务器上或者另一台没有开发环境的计算机上。部署工作分成两部分，一是数据库的迁移；二是项目本身的异机部署。要注意服务器上的运行环境，如数据库的端口和密码、Tomcat 或其他容器的端口、防火墙的设置等，如果与开发机器有区别，可能涉及到修改配置文件。

只有把自己或团队合作开发的项目部署到服务器，经过远程测试，一切正常，才能说明完整地掌握了 JavaEE 项目的一般开发和部署流程。

参考文献

[1] 赵彦．JavaEE 框架技术进阶式教程[M]．北京：清华大学出版社，2011．

[2] 郭克华．JavaEE 程序设计与应用开发[M]．北京：清华大学出版社，2011．

[3] 张继军，董卫. Java EE 框架开发技术与案例教程[M]. 北京：机械工业出版社，2016.

[4] 黄玲，罗丽娟. JavaEE 程序设计及项目开发教程[M]. 重庆：重庆大学出版社，2017.

[5] zozoh. Nutz 使用手册[EB/OL]. [2017-7-17]. https://nutzam.com/core/nutz_preface.html.

[6] 李家智．Beetl 文档[EB/OL]．[2018]．http://ibeetl.com/guide/#beetl．

[7] FEX 前端研发团队．UEditor 文档[EB/OL]．[2016]．http://fex.baidu.com/ueditor/.

[8] 詹波．JFinal 文档[EB/OL]．[2016]．http://www.jfinal.com/doc．

[9] EasyUI 中文小组．Jquery EasyUI 中文网[EB/OL]．[2015]．http://www.jeasyui.net/.

附录　在线资源

JDK（Java Development Kit，http://www.oracle.com/technetwork/java/javase/downloads/）是Java 语言的软件开发工具包（SDK），它是整个 Java 开发的核心，包含了 Java 的运行环境 JRE（JVM+Java 系统类库）和 Java 工具。

Tomcat（http://tomcat.apache.org/）是 Apache 的 Jakarta 项目中的一个核心项目，由于技术先进、性能稳定、免费开源，已成为比较流行的 Web 应用服务器。

MySQL（https://dev.mysql.com/downloads/）是最流行的关系型数据库管理系统之一，在Web 应用方面，MySQL 是最好的 RDBMS 应用软件之一。

MyEclipse（http://www.myeclipsecn.com/download/）是功能丰富的 JavaEE 集成开发环境，包括了完备的编码、调试、测试和发布功能，完整支持 HTML、CSS、Javascript、JSP、SQL、Struts、Spring 和 Hibernate 等。

HeidiSQL（https://www.heidisql.com/download.php）是一款操作简单的 MySQL 数据库可视化管理工具。

Notepad++（https://notepad-plus-plus.org/）是一款非常有特色的编辑器，是开源软件，可以免费使用。Notepad++功能比 Windows 中的 Notepad（记事本）强大，除了可以用来制作一般的纯文字说明文件，也十分适合编写计算机程序代码。Notepad++不仅有语法高亮度显示功能，也有语法折叠功能，并且支持宏以及扩充基本功能的外挂模组。

MagicTools（http://otom31.iteye.com/blog/2408835）是采用逆向工程，根据数据库自动生成代码的工具，支持模板，可以直接生成 entity/pojo，dao，controller，service 等类型的代码。

Nutz（https://nutzam.com/core/nutz_preface.html）是永久免费、完整开源、体积小巧无依赖、功能强大、开发速度快的 JavaEE 项目开发框架。

NutzWk（https://wizzer.cn/）是基于 Nutz 的开源企业级开发框架，模块化、接口化、统一提供代码生成器、IDEA 插件等。

Struts2（https://struts.apache.org/）是一个基于 MVC 设计模式的 Web 应用框架，它本质上相当于一个 servlet，在 MVC 设计模式中，Struts2 作为控制器（Controller）来建立模型与视图的数据交互。Struts 2 以 WebWork 为核心，采用拦截器的机制来处理用户的请求，使得业务逻辑控制器能够与 ServletAPI 完全脱离开。

Spring（https://spring.io/）是一个轻量级的 Java 开发框架，是一个开放源代码的设计层面框架，解决业务逻辑层和其他各层的松耦合问题。它将面向接口的编程思想贯穿整个系统应用。

Spring MVC（https://docs.spring.io/spring/docs/current/spring-framework-reference/web.html）是 Spring 框架最重要的的模块之一。它以强大的 Spring IoC 容器为基础，并充分利用容器的特性来简化它的配置。Spring MVC 框架分离了控制器、模型对象、过滤器以及处理程序对象的角色，这种分离更容易进行定制。

Hibernate（http://hibernate.org/）是一个基于 jdbc 的开源的持久化框架，是一个优秀的开放源代码的对象关系映射 ORM 框架，很大程度地简化了 DAO 层编码工作。它对 JDBC 进行了非常轻量级的对象封装，它将 Pojo 与数据库表建立映射关系，可以自动生成 SQL 语句，自动执行，使得 Java 程序员可以随心所欲地使用对象编程思维来操纵数据库。Hibernate 可以应用在任何使用 JDBC 的场合，既可以在 Java 的客户端程序使用，也可以在 Servlet/JSP 的 Web 应用中使用。

MyBatis（http://www.mybatis.org/）的前身是 iBatis，来源于“internet”和“abatis”的组合。MyBatis 是一款优秀的持久层框架。iBATIS 提供的持久层框架包括 SQLMaps 和 Data Access Objects（DAOs），它支持定制化 SQL、存储过程以及高级映射。MyBatis 避免了几乎所有的 JDBC 代码和手动设置参数以及获取结果集。MyBatis 可以使用简单的 XML 或注解来配置和映射原生信息，将接口和 Java 的 Pojos 映射成数据库中的记录。

JFinal（http://www.jfinal.com/）是基于 Java 语言的极速 Web + ORM 框架，其核心设计目标是开发迅速、代码量少、学习简单、功能强大、轻量级、易扩展、Restful。在拥有 Java 语言所有优势的同时再拥有 ruby、python、php 等动态语言的开发效率。

EasyUI（http://www.jeasyui.net/）基于 jQuery 的用户界面插件集合，包括强大的 DataGrid，树网格，面板等，完美支持 HTML5 网页，使用 EasyUI 节省网页开发的时间和规模，只需要通过编写一些简单 HTML 标记就可以定义用户界面。

w3school（http://www.w3school.com.cn/）领先 Web 技术教程，全部免费，从基础的 HTML 到 CSS，乃至进阶的 HTML5、CSS3、XML、SQL、JS、PHP 和 ASP.NET。

菜鸟教程（http://www.runoob.com/）提供了最全的基础编程技术教程，包括了 HTML、CSS、Javascript、PHP、C、Python 等各种基础编程教程；同时提供了大量的在线实例，通过实例，可以更好地学习如何建站；所有资源是完全免费的，并且会根据当前互联网的变化实时更新本站内容。

UEditor（http://ueditor.baidu.com/，https://github.com/fex-team/ueditor）是一套开源的在线 HTML 编辑器，主要用于让用户在网站上获得所见即所得编辑效果。开发人员可以用 UEditor 把传统的多行文本输入框（textarea）替换为可视化的富文本输入框。UEditor 使用 JavaScript 编写，可以无缝地与 Java、.NET、PHP、ASP 等程序集成，比较适合在 CMS、商城、论坛、博客、Wiki、电子邮件等互联网应用上使用。

Beetl（Bee Template Language，http://ibeetl.com/）是新一代的 Java 模板引擎，功能齐全，语法直观，模板易维护，性能远远超过当前流行的其他模板引擎，语法类似 Javascript，易学易用。

Maven 中央存储库（http://mvnrepository.com/）是 Maven 用来下载所有项目的依赖库，里面有大量的常用类库，并包含了世界上大部分流行的开源项目构件，目前是以 Java 为主工程依赖的 Jar 包。

ZBUS（https://www.oschina.net/p/zbus）是高可用消息队列框架（ZBUS = MQ + RPC + PROXY），支持消息队列、发布订阅、RPC、代理（TCP/HTTP/DMZ）亿级消息堆积能力、支持 HA 高可用超轻量级、单个 Jar 包无依赖。

store（https://github.com/marcuswestin/store.js）实现了浏览器的本地存储的 JavaScript 封装 API，不是通过 Cookie 和 Flash 技术实现，而是使用 localStorage、globalStorage 和 userData 行为。

zTree（http://www.treejs.cn/v3/main.php）是一个依靠 jQuery 实现的开源免费的多功能“树插件”。优异的性能、灵活的配置和多种功能的组合是 zTree 最大优点。

BootCDN（http://www.bootcdn.cn/）是 Bootstrap 中文网和又拍云共同支持并维护的前端开源项目免费 CDN 服务，由又拍云提供全部 CDN 支持，致力于为 Bootstrap、jQuery、Angular 一类优秀的前端开源项目提供稳定、快速的免费 CDN 加速服务。BootCDN 所收录的开源项目主要同步于 cdnjs 仓库。

POI（http://poi.apache.org/）是 Apache 软件基金会的开放源码函数库，POI 提供 API 给 Java 程序对 Microsoft Office 格式文件读和写的功能。其中 HSSF 提供读写 Microsoft Excel 格式文件的功能；HWPF 提供读写 Microsoft Word 格式档案的功能；HSLF 提供读写 Microsoft PowerPoint 格式文件的功能。

GitHub（https://github.com）是一个面向开源及私有软件项目的托管平台，具有优越的版本控制系统，可以作为免费的远程源码仓库，还是一个开源协作社区。通过 GitHub 既可以让别人参与你的开源项目，你也可以参与别人的开源项目。

码云（https://gitee.com/）专为开发者提供稳定、高效、安全的云端软件开发协作平台，是开源中国推出的代码托管平台，支持 Git 和 SVN，提供免费的私有仓库托管。无论是个人、团队、企业都能够用码云实现代码托管、项目管理、协作开发。

Font Awesome（http://fontawesome.dashgame.com/）是一套绝佳的图标字体库和 CSS 框架，提供可缩放的矢量图标，可以使用 CSS 控制图标的大小、颜色、阴影或者其他任何支持的效果。

Bootstrap（http://www.bootcss.com/）是一款简洁、直观、强悍的前端开发框架，可以让Web 开发更迅速、简单。

模板之家（http://www.cssmoban.com/）力争为大家提供最好最全的网站模板、DIV+CSS 模板、Wordpress 主题模板、CSS Menu 等实用资源。网页模板都是站长从国外大小网站收集而来，旨在为开发者在工作或学习时提高效率、节省时间。